MULTICOMPONENT POLYMERIC MATERIALS

From Introduction to Application

MULTICOMPONENT POLYMERIC MATERIALS

From Introduction to Application

Edited by
Gennady E. Zaikov, DSc, Nodar G. Lekishvili, DSc, and Yurii G. Medvedevskikh, DSc

Apple Academic Press

TORONTO   NEW JERSEY

© 2013 by
Apple Academic Press Inc.
3333 Mistwell Crescent
Oakville, ON L6L 0A2
Canada

Apple Academic Press Inc.
1613 Beaver Dam Road, Suite # 104
Point Pleasant, NJ 08742
USA

First issued in paperback 2021

Exclusive worldwide distribution by CRC Press, a Taylor & Francis Group

ISBN 13: 978-1-77463-267-3 (pbk)
ISBN 13: 978-1-926895-35-2 (hbk)

Library of Congress Control Number: 2012951947

Library and Archives Canada Cataloguing in Publication

Multicomponent polymeric materials: from introduction to application/edited by Gennady E. Zaikov, Nodar G. Lekishvili, and Yurii J. Medvedevskikh.

Includes bibliographical references and index.
ISBN 978-1-926895-35-2
1. Polymers. 2. Materials science. 3. Chemical engineering. I. Zaikov,
G. E. (Gennadi˘i Efremovich), 1935- II. Lekishvili, N III. Medvedevskikh, Y. G

QD381.M85 2013   547'.7   C2012-906366-5

About the Editors

Gennady E. Zaikov, DSc

Gennady E. Zaikov, DSc, is Head of the Polymer Division at the N. M. Emanuel Institute of Biochemical Physics, Russian Academy of Sciences, Moscow, Russia, and Professor at Moscow State Academy of Fine Chemical Technology, Russia, as well as Professor at Kazan National Research Technological University, Kazan, Russia. He is also a prolific author, researcher, and lecturer. He has received several awards for his work, including the the Russian Federation Scholarship for Outstanding Scientists. He has been a member of many professional organizations and on the editorial boards of many international science journals.

Nodar G. Lekishvili, DSc

Dr. Nodar Lekishvili is Professor and Chair of the General, Inorganic and Organometallic Chemistry Department and the Director of the Institute of Inorganic-Organic Hybrid Compounds and Nontraditional Materials at Tbilisi State University (TSU), Georgia. He received is PhD from the Institute of the Organometallic Chemistry RAS (Moscow). His scientific interests include the coordination of compounds with element organic and organic ligands; compounds and materials with specific properties; element (metal) organic compounds with nontraditional structures and properties; inorganic-organic hybrid materials; and composite materials with specific properties. He has written more than 250 scientific publications, including monographs, textbooks, and three books and has several inventions and patents to his name.

Yurii G. Medvedevskikh, DSc

Yurii G. Medvedevskikh is Chief Scientist at the Department of Physico-Chemistry and Technology of Combustible Minerals at the L. M. Lytvynenko Institute of Physico-Organic Chemistry and Coal Chemistry of the National Academy of Sciences of Ukraine, in Lviv, Ukraine. He is a scientist in the field of chemistry and the physics of oligomers, polymers, composites, and nanocomposites.

Contents

List of Contributors

O. V. Afonicheva
Nesmeyanov Institute of Organoelement Compounds, Russian Academy of Sciences, 28 Vavilov St., Moscow-119991, Russia.

Yu. O. Andriasyan
Institutenstitute of Biochemical Physics, Russian Academy of Sciences, Moscow, Russia.

D. A. Beeva
Kabardino-Balkarian State University, Chernyshevsky st., 173, Nal'chik-360004, Russian Federation.

A. L. Belousova
Institutenstitute of Biochemical Physics, Russian Academy of Sciences, Moscow, Russia.
Moscow State Academy of Fine Chemical Technology, Moscow, Russia.

O. I. Demchyna
Department of Physico-chemistry of Combustible Minerals L.M. Lytvynenko Institute of Physico-Organic Chemistry and Coal Chemistry NAS of Ukraine Naukova St., 3a, Lviv-79059, Ukraine.

L. S. Galbraikh
Moscow State Textile University, Malaya Kaluzhskaya ul. 1, Moscow-119071 Russia.

A. R. Khokhlov
Nesmeyanov Institute of Organoelement Compounds, Russian Academy of Sciences, 28 Vavilov St., Moscow-119991, Russia.

N. R. Kildeeva
Moscow State Textile University, Malaya Kaluzhskaya ul. 1, Moscow-119071 Russia.

V. V. Kochubey
Lviv Polytechnic National University Bandera St., 12, Lviv-79013, Ukraine.

A. E. Kornev
Institutenstitute of Biochemical Physics, Russian Academy of Sciences, Moscow, Russia
Moscow State Academy of Fine Chemical Technology, Moscow, Russia.

A. N. Kosygin
Moscow State Textile University, Malaya Kaluzhskaya ul. 1, Moscow-119071 Russia.

Z. M. Koval'
Lviv Polytechnic National University Bandera St., 12, Lviv-79013, Ukraine.

G. V. Kozlov
Institute of Applied Mechanics of Russian Academy of Sciences, Leninskii pr., 32 A, Moscow-119991, Russian Federation.

A. P. Krasnov
Nesmeyanov Institute of Organoelement Compounds, Russian Academy of Sciences, 28 Vavilov St., Moscow-119991, Russia.

Stefan Kubica
Institut Inzynierii Materialow Polimerowych I Barwnikow, 55 M. Sklodowskiej-Curie str., 87-100 Torun, Poland.

Nodar Lekishvili
Ivane Javakhishvili Tbilisi State University, Faculty of Exact and Natural Sciences 1, Ilia Chavchavadze Ave., 0128 Tbilisi, Georgia.

Yu. G. Medvedevskikh
Department of Physico-chemistry of Combustible Minerals L. M. Lytvynenko Institute of Physico-Organic Chemistry and Coal Chemistry NAS of Ukraine Naukova St., 3a, Lviv-79059, Ukraine.

I. A. Mikhaylov
Institutenstitute of Biochemical Physics, Russian Academy of Sciences, Moscow, Russia.

A. K. Mikitaev
Kabardino-Balkarian State University, Chernyshevsky st., 173, Nal'chik-360004, Russian Federation.

V. A. Mit'
Nesmeyanov Institute of Organoelement Compounds, Russian Academy of Sciences, 28 Vavilov St., Moscow-119991, Russia.

Yu. G. Moskalev
Institutenstitute of Biochemical Physics, Russian Academy of Sciences, Moscow, Russia "Polykrov", Moscow, Russia.

Levan Nadareishvili
Cybernetics Institute, 5, S. Euli, str., 0186 Tbilisi, Georgia.

A. V. Naumkin
Nesmeyanov Institute of Organoelement Compounds, Russian Academy of Sciences, 28 Vavilov St., Moscow-119991, Russia.

A. Y. Nikolaev
Nesmeyanov Institute of Organoelement Compounds, Russian Academy of Sciences, 28 Vavilov St., 119991 Moscow, Russia.

A. A. Panov
Sterlitamak Institute for Applied Research of Academy of Sciences, Republic of Bashkortostan, Sterlitamak-453103, Odesskaya Street, 68.

A. K. Panov
Sterlitamak Institute for Applied Research of Academy of Sciences, Republic of Bashkortostan, Sterlitamak-453103, Odesskaya Street, 68.

A. A. Popov
Institutenstitute of Biochemical Physics, Russian Academy of Sciences, Moscow, Russia.

H. V. Romaniuk
Lviv Polytechnic National University Bandera St., 12, Lviv-79013, Ukraine.

S. N. Rusanova
Kazan State Technological University, K.Marx, 68, Kazan, Tatarstan, Russia-420015.

E. E. Said-Galiev
Nesmeyanov Institute of Organoelement Compounds, Russian Academy of Sciences, 28 Vavilov St., Moscow-119991, Russia.

S. Ju. Sofina
Kazan State Technological University, K.Marx, 68, Kazan, Tatarstan-420015, Russia.

A. N. Sonina
Moscow State Textile University, Malaya Kaluzhskaya ul. 1, Moscow-119071 Russia.

O. V. Stoyanov
Kazan State Technological University, K.Marx, 68, Kazan, Tatarstan-420015, Russia.

Nona Topuridze
Cybernetics Institute, 5, S. Euli, str., 0186 Tbilisi, Georgia.

S. A. Uspenskiy
Moscow State Textile University, Malaya Kaluzhskaya ul. 1, Moscow-119071 Russia.

G. A. Vikhoreva
Moscow State Textile University, Malaya Kaluzhskaya ul. 1, Moscow-119071 Russia.

I. O. Volkov
Nesmeyanov Institute of Organoelement Compounds, Russian Academy of Sciences, 28 Vavilov St., Moscow-119991, Russia.

Zurab Wardosanidze
Cybernetics Institute, 5, S. Euli, str., 0186 Tbilisi, Georgia.

Yu. G. Yanovskii
Institute of Applied Mechanics of Russian Academy of Sciences, Leninskii pr., 32 A, Moscow-119991, Russian Federation.

I.Yu Yevchuk
Department of Physico-chemistry of Combustible Minerals L. M. Lytvynenko Institute of Physico-Organic Chemistry and Coal Chemistry NAS of Ukraine Naukova St., 3a, Lviv-79059, Ukraine.

G. E. Zaikov
Sterlitamak Institute for Applied Research of Academy of Sciences, Republic of Bashkortostan, Sterlitamak-453103, Odesskaya Street, 68.
Institute of Biochemical Physics n.a. N.M. Emanuel, Russian Academy of Sciences, Moscow-119991, Kosygin Street, 4.

List of Abbreviations

ACA	ε-Amino caproic acid
ASF	Atomic sensitivity factors
ATR	Attenuated total reflection
BBR	Bromated butyl rubber
B-PES	Bromide-containing aromatic copolyethersulfones
BSR	Butadiene styrene rubber
CBR	Chlorinated butyl rubber
CEP	Chlorinated ethylene-propylene
CEPDC	Chemical Engineering Process Design Centre
CEPDC	Chlorine-containing ethylene-propylene-diene cauotchoucs
CMC	Cell membrane complex
CO	Carbon oxide
CP	Chlorinated polyethylene
CSP	Chlorosulfonated polyethylene
EPDC	Ethylene-propylene-diene cauotchoucs
EVA	Ethylene vinyl acetate
EVAMA	Vinyl acetate and maleic anhydride
GB	Gradient birefringence
GBFS	Granulated blast-furnace slag
GRIN	Gradient Refractive Index
HM	Halide modification
IR	Infrared
ISIRI	Institute of Standards and Industrial Research of Iran
LMC	Low molecular compound
MFR	Melt flow rate
NR	Natural rubber
OPC	Ordinary portland cement
PC	Polymerizing composition
PCA	Polycaproamide
PHE	Poly(hydroxyl ether)
PPX	Poly(phenylxalines)
PVA	Poly(vinyl alcohol)
PVDF	Poly(vinylidene fluoride)
SC-CO$_2$	Supercritical carbon dioxide
Selfoc	Self focusing
SEM	Scanning electron micrographs
SGS	Sol-gel system
SIR-3	Synthetic isoprene rubber
SPIP	Scanning probe image processor
TC	Technical carbon
TEM	Transmissionelectron microscopy

TMDA	Tetramethylene acrylate
UHMWPE	Ultra-high-molecular-weight polyethylene
WPLA	Waste PET bottles lightweight aggregate
WPLAC	Waste PET bottles lightweight aggregate concrete

Preface

The book aimed at giving a detailed overview of the main and most up-to-date advances in the area of polymeric materials, through a balanced combination of theory and experiments. Since the subject is essentially an interdisciplinary area and as such it brings together scientists and engineers with different educational backgrounds, it was important to offer a research-oriented exposition of the fundamentals as well.

This book is based on the editors' extensive experience in research, development, and education in the field of materials science and especially polymer testing, polymer diagnostics, and failure analysis. The results of their work were published in several reference books about deformation and fracture behavior of polymers, in numerous single publications in peer-reviewed scientific journals, and in proceedings. Given the fact that the field of science undergoes a rapid and dynamic development, it seemed prudent to present these results here.

The book presents a comprehensive representation of knowledge provided by respected colleagues from universities and from the polymer industry.

This book is primarily designed for students of bachelor, diploma and master courses of materials science, materials technology, plastic technology, mechanical engineering, process engineering, and chemical engineering. It can be used by students, teachers of universities and colleges for supplementary studies in the disciplines of chemistry and industrial engineering. The methods of polymer testing are also essential to the development and application of biomedical or nanostructured materials. With the publication of this book we hope that it will not only serve the important task of training of young scientists in physical and materials oriented disciplines, but will also make a contribution to further the education of professional polymer testers, design engineers, and technologists.

— **Gennady E. Zaikov, DSc**

1 Updates on Nanofiller Structure in Elastomeric Nanocomposites

G. V. Kozlov, Yu. G. Yanovskii, S. Kubica, and G. E. Zaikov

CONTENTS

1.1 INTRODUCTION

It is well known fact that, in particulate-filled elastomeric nanocomposites (rubbers) nanofiller particles form linear spatial structures ("chains") [1, 2]. At the same time in polymer composites, filled with disperse microparticles (microcomposites) and particles (aggregates of particles) of filler form a fractal network, that defines polymer matrix structure (analog of fractal lattice in computer simulation) [3], which results in different mechanisms of polymer matrix structure, forms micro and nanocomposites. If the first filler particles (aggregates of particles), fractal network availability results to "disturbance" of polymer matrix structure, expressed in the increase in its fractal dimension d_f [3], then in case of polymer nanocomposites at nanofiller contents change, the value d_f is not changed and equal to matrix polymer structure fractal dimension [4]. It has to been expected composites, indicated classes structure formation mechanism change defines their properties change, in particular, reinforcement degree.

At present, there are several methods of filler structure (distribution) determination in polymer matrix, both experimental [5, 6] and theoretical [3]. All the indicated methods describe this distribution by fractal dimension D_n of filler particles network. However, correct determination of any object fractal (Hausdorff) dimension includes three obligatory conditions. The first is the indicated above determination of fractal dimension numerical magnitude, which should not be equal to object topological dimension. Any real (physical) fractal possesses [7] fractal properties within a certain scales range [8]. And at last, the third condition is the correct choice of measurement scales range itself. As it has been shown [9, 10], the minimum range should exceed at any rate self-similarity iteration.

The purpose of present chapter is dimension D_n estimation, both experimentally and theoretically, and to check, two indicated conditions to fulfill. That is to obtain of nanofiller particles (aggregates of particles) network ("chains") fractality strict proof in elastomeric nanocomposites on the example of particulate-filled butadiene styrene rubber (BSR).

1.2   EXPERIMENTAL

The elastomeric particulate-filled nanocomposite on the basis of BSR was an object of the study. The technical carbon of mark № 220 (TC) of industrial production, nano and microshungite (the mean filler particles size makes up 20, 40, and 200 nm, respectively) were used as a filler. All fillers content makes up 37 mass%. Nano and microdimensional disperse shungite particles were obtained from industrially extractive material by processing according to the original technology. Size and polydispersity of shungite particles received in milling process were monitored with the aid of analytical disk centrifuge (CPS Instruments, Inc., USA), allowing to determine with high precision the size and distribution by sizes within the range from 2 nm to 50 mcm.

Nanostructure was studied on atomic power microscopes Nano-DST (Pacific Nanotechnology, USA) and Easy Scan DFM (Nanosurf, Switzerland) by semi-contact method in the force modulation regime. Atomic power microscopy results were processed with the aid of specialized software package Scanning Probe Image Processor, Denmark (SPIP). The SPIP is a powerful program package for processing of images, obtained on SPM, AFM, STM, scanning electron microscopes, transmission electron microscopes, interferometers, confocal microscopes, profilometers, optical microscopes, and so on.

The given package possesses the whole functions numbers that are necessary for images precise analysis, included as follows:

1. The possibility of three-dimensional reflecting objects obtaining, distortions automatized leveling, including Z-error mistakes removal for examination separate elements, and so on;
2. Quantitative analysis of particles or grains, more than 40 parameters can be calculated for each found particle or pore-area, perimeter, average diameter, the ratio of linear sizes of grain width to its height distance between grains, coordinates of grain center of mass a.a. can be presented in a diagram form or in a histogram form.

1.3   DISCUSSION AND RESULTS

The first method of dimension D_n experimental determination uses the fractal relationship [11, 12] as follows:

$$D_n = \frac{\ln N}{\ln \rho} \tag{1}$$

where N is a number of particles with size ρ.

Particles sizes were established on the basis of atomic power microscopy data (see Figure 1). For each from the three studied nanocomposites no less than 200 particles were measured, the sizes of which were united into 10 groups and mean values N and ρ were obtained. The dependences $N(\rho)$ in double logarithmic coordinates were plotted, which proved to be linear and the values D_n were calculated according to their slope (see Figure 2). It is obvious, that at such approach fractal dimension D_n is determined in two-dimensional Euclidean space, whereas real nanocomposite should be considered in three-dimensional Euclidean space. The following relationship can be used for D_n re-calculation for the case of three-dimensional space [13]:

$$D3 = \frac{d + D2 \pm \left[(d - D2)^2 - 2 \right]^{1/2}}{2} \tag{2}$$

where $D3$ and $D2$ are corresponding fractal dimensions in three- and two-dimensional Euclidean spaces, $d = 3$.

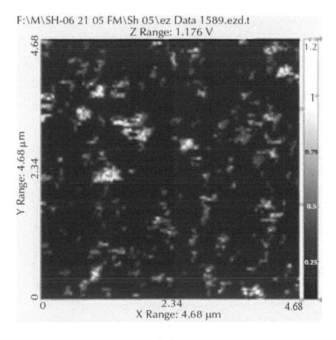

(a)

FIGURE 1   *(Continued)*

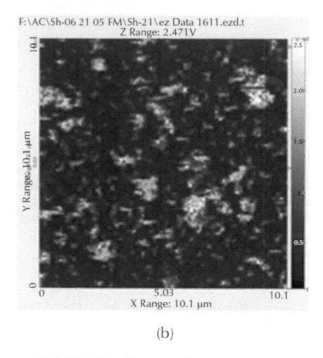

(b)

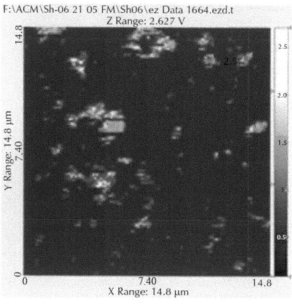

(C)

FIGURE 1 The electron micrographs of nanocomposites BSR/TC (a), BSR/nanoshungite (b), and BSR/microshungite (c), obtained by atomic power microscopy in the force modulation regime.

It calculated according to the indicated method dimensions D_n are adduced in Table 1. As it follows from the data of this table, the values D_n for the studied nanocomposites are varied within the range of 1.10-1.36, that is they characterize more or less branched linear formations ("chains") of nanofiller particles (aggregates of particles) in elastomeric nanocomposite structure. Let us remind that for particulate-filled composites poly(hydroxiether)/graphite the value D_n changes within the range of ~2.30-2.80 [5], that is for these materials filler particles network is a bulk object, but not a linear one [7].

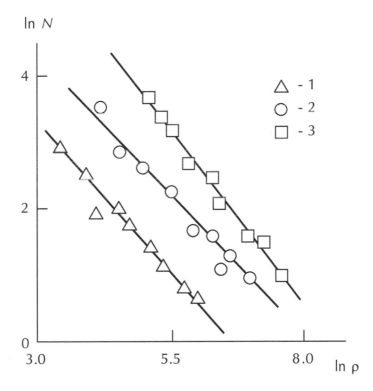

FIGURE 2   The dependence of nanofiller particles number N on their size ρ for nanocomposites BSR/TC (1), BSR/nanoshungite (2), and BSR/microshungite (3).

TABLE 1   The dimensions of nanofiller particles (aggregates of particles) structure in elastomeric nanocomposites.

The nanocomposite	D_n, the equations (1)	D_n, the equations (3)	d_0	d_{surf}	φ_n	D_n, the equations (7)
BSR/TC	1.19	1.17	2.86	2.64	0.48	1.11
BSR/nanoshungite	1.10	1.10	2.81	2.56	0.36	0.78
BSR/microshungite	1.36	1.39	2.41	2.39	0.32	1.47

Another method of D_n experimental determination uses the so called "quadrates method" [14]. Its essence consists in the following—On the enlarged nanocomposite microphotograph (see Figure 1) a net of quadrates with side size α_i, changing from 4.5 to 24 mm with constant ratio $\alpha_{i+1}/\alpha_i = 1.5$, is applied and then quadrates number N_i, in to which nanofiller particles hit (fully or partly), is calculated. Five arbitrary net positions concerning microphotograph were chosen for each measurement. If nanofiller particles network is fractal, then the following relationship should be fulfilled [14]:

$$N_i \sim S_i^{-D_n/2} \tag{3}$$

where S_i is quadrate area, which is equal to α_i^2.

In Figure 3 the dependence of N_i on S_i in double logarithmic coordinates for the three studied nanocomposites, corresponding to the relationship (3), is adduced. As one can see, these dependences are linear, that allows to determine the value D_n from

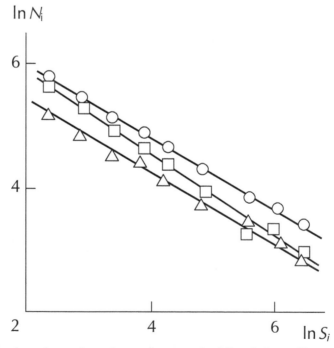

FIGURE 3 The dependence of covering quadrates number Ni on their area Si, corresponding to the relationship (3), in double logarithmic coordinates for nanocomposites on the basis of BSR. The designations are the same, that in Figure 2.

their slope. It determined according to the relationship (3) values D_n that are also adduced in Table 1, from which a good correspondence of dimensions D_n, obtained by the above two described methods, follows (their average discrepancy makes up 2.1% after these dimensions recalculation for three-dimensional space according to the Equation (2)).

As it has been shown [15], at the relationship (3) the usage for self-similar fractal objects the condition should be fulfilled:

$$N_i - N_{i-1} \sim S_i^{-D_n} \qquad (4)$$

In Figure 4 the dependence, corresponding to the relationship (4), for the three studied elastomeric nanocomposites is adduced. As one can see, this dependence is linear, passes through coordinates origin that according to the relationship (4) is confirmed by nanofiller particles (aggregates of particles) "chains" self-similarity within the selected α_i range. It is apparent, that this self-similarity will be a statistical one [15]. Let us note, that the points, corresponding to $\alpha_i = 16$ mm for nanocomposites BSR/TC and BSR/microshungite, do not correspond to a common straight line. Accounting for electron microphotographs of Figure 1 enlargement gives the self-similarity range for nanofiller "chains" of 464–1472 nm. For nanocomposite BSR/nanoshungite, which has no points deviating from a straight line of Figure 4, α_i range makes up 311–1510 nm, that corresponds well enough to the indicated above self-similarity range.

It has been shown [9, 10], that measurement scales S_i minimum range should contained at least one self-similarity iteration. In this case the condition for ratio of maximum S_{max} and minimum S_{min} areas of covering quadrates should be fulfilled [10]:

$$\frac{S_{max}}{S_{min}} > 2^{2/D_n} \qquad (5)$$

Hence, accounting for the defined above restriction, let us obtain $S_{max}/S_{min} = 121/20.25 = 5.975$, that is larger than values $2^{2/D_n}$ for the studied nanocomposites, which are equal to 2.71–3.52. This means, that measurement scales range is chosen correctly.

The self-similarity iterations number μ can be estimated from the inequality [10]:

$$\left(\frac{S_{max}}{S_{min}}\right)^{D_n/2} > 2^{\mu} \qquad (6)$$

Using the indicated above values of the included in the inequality (6) parameters, $\mu = 1.42$-1.75 is obtained for the studied nanocomposites, that is in our experiment conditions self-similarity iterations number is larger than unity, that again is confirmed by the value D_n estimation correctness [6].

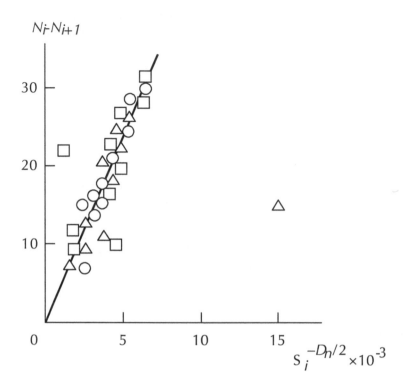

FIGURE 4 The dependences of (Ni–Ni+1) on the value $S_i^{-D_n/2}$, corresponding to the relationship (4), for nanocomposites on the basis of BSR. The designations are the same, that in Figure 2.

And let us consider in conclusion the physical grounds of smaller values D_n for elastomeric nanocomposites in comparison with polymer microcomposites, that is, the causes of nanofiller particles (aggregates of particles) "chains" formation in the first. The value D_n can be determined theoretically according to the equation [3]:

$$\varphi_{if} = \frac{D_n + 2.55d_0 - 7.10}{4.18} \tag{7}$$

where φ_{if} is interfacial regions relative fraction, d_0 is nanofiller initial particles surface dimension.

The dimension d_0 estimation can be carried out with the aid of the relationship [4]:

$$S_u = 410\left(\frac{D_p}{2}\right)^{d_0-d} \tag{8}$$

where S_u is nanofiller initial particles specific surface in m²/g, D_p is their diameter in nm, d is dimension of Euclidean space, in which a fractal is considered (it is obvious, in our case $d = 3$).

The value S_u can be calculated according to the equation [16]:

$$S_u = \frac{6}{\rho_n D_p} \qquad (9)$$

where ρ_n is nanofiller density, which is determined according to the empirical formula [4]:

$$\rho_n = 0.188(D_p)^{1/3} \qquad (10)$$

The results of value d_0 theoretical estimation are adduced in Table 1. The value φ_{if} can be calculated according to the equation [4]:

$$\varphi_{if} = \varphi_n (d_{surf} - 2) \qquad (11)$$

where φ_n is nanofiller volume fraction, d_{surf} is fractal dimension of nanoparticles aggregate surface.

The value φ_n is determined according to the equation [4]:

$$\varphi_n = \frac{W_n}{\rho_n} \qquad (12)$$

where W_n is nanofiller mass fraction and dimension d_{surf} is calculated according to the Equations (8)-(10) at diameter D_p replacement on nanoparticles aggregate diameter D_{agr}, which is determined experimentally (see Figure 5).

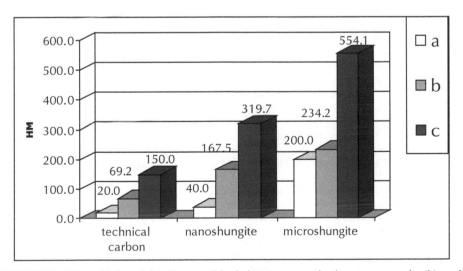

FIGURE 5 The initial particles diameter (a), their aggregates size in nanocomposite (b), and distance between nanoparticles aggregates (c) for nanocomposites on the basis of BSR, filled with technical carbon, nano and microshungite.

The results of dimension D_n theoretical calculation according to the Equations (7)-(12) are adduced in Table 1, which shows good correspondence of theory and experiment. The Equation (7) indicates unequivocally the cause of filler in nano and microcomposites different behavior. The high (close to 3, see Table 1) values d_0 for nanoparticles and relatively small ($d_0 = 2.17$ for graphite) values d_0 for microparticles at comparable values φ_{if} for composites of the indicated classes [3, 4].

1.4 CONCLUSION

Therefore, the results have shown, that nanofiller particles (aggregates of particles) "chains" in elastomeric nanocomposites are physical fractal within self-similarity (and, hence, fractality [12]) range of ~500-1450 nm. In this range their dimension D_n can be estimated according to the Equations (1), (3), and (7). The cited examples demonstrate the necessity of the measurement scales range correct choice. As it has been noted earlier [17], linearity of the plots, corresponding to the Equations (1) and (3), and D_n nonintegral value do not guarantee object self-similarity (and hence, fractality). The nanofiller particles (aggregates of particles) structure low dimensions are due to the initial nanofiller particles surface high fractal dimension.

KEYWORDS

- **Atomic power microscopy data**
- **Butadiene-styrene rubber**
- **Euclidean space**
- **Nanocomposites**

REFERENCES

1. Lipatov, Yu. S. *The Physical Chemistry of Filled Polymers*. Moscow, Chemistry Publishing House, p. 304 (1977).
2. Bartenev, G. M. and Zelenev, Yu. V. *The Physics and Mechanics of Polymers*. Moscow, Vysshaya Shkola Publishing House, p. 391 (1983).
3. Kozlov, G. V., Yanovskii, Yu. G., and Zaikov, G. E. Structure and Properties of Particulate-Filled Polymer Composites: *The Fractal Analysis*. Nova Science Publishers, Inc., New York, p. 282 (2010).
4. Mikitaev, A. K., Kozlov, G. V., and Zaikov, G. E. Polymer Nanocomposites: *The Variety of Structural Forms and Applications*. Nova Science Publishers, Inc., New York, p. 319 (2008).
5. Kozlov, G. V. and Mikitaev, A. K. Mechanics of Composite Materials and Constructions (in rus.), 2(3–4), 144–157 (1996).
6. Kozlov, G. V., Yanovskii, Yu. G., and Mikitaev, A. K. Mechanics of Composite Materials and Constructions (in rus.), 34(4), 539–544 (1998).
7. Balankin, A. S. *Synergetics of Deformable Body*. Moscow, Publishers of Ministry Defence USSR, p. 404 (1991).
8. Hornbogen, E. *Intern. Mater. Rev.*, 34(6), 277–296 (1989).
9. Pfeifer, P. *Appl. Surf. Sci.*, 18(1), 146–164 (1984).
10. Avnir, D., Farin, D., and Pfeifer, P. *J. Colloid Interface Sci.*, 103(1), 112–123 (1985).
11. Ishikawa, K. *J. Mater. Sci. Lett.*, 9(4), 400–402 (1990).

12. Ivanova, V. S., Balankin, A. S., Bunin, I. Zh., and Oksogoev, A. A. *Synergetics and Fractals in Material Science*. Moscow, Nauka Publishing House, p. 383 (1994).
13. Vstovskii, G. V., Kolmakov, L. G., and Terent'ev, V. F. *Metals (in rus.)*, (4), 164–178 (1993).
14. Hansen J. P. and Skjeitorp A. T. *Phys. Rev. B*, **38**(4), 2635–2638 (1988).
15. Pfeifer, P., Avnir, D., and Farin, D. *J. Stat. Phys.*, **36** (5/6), 699–716 (1984).
16. Bobryshev, A. N., Kozomazov, V. N., Babin, L. O., and Solomatov, V. I. *Synergetics of Composite Materials,* Lipetsk-city, NPO ORIUS Publishing House, p. 154 (1994).
17. Farin, D., Peleg, S., Yavin, D., and Avnir, D. *Langmuir*, **1**(4), 399–407 (1985).

2 Fractal Clusters in Physics-Chemistry of Polymers

G. V. Kozlov, D. A. Beeva, G. E. Zaikov, A. K. Mikitaev, and S. Kubica

CONTENTS

2.1 INTRODUCTION

In the last 25 years an interest of physicists to the theory of polymer synthesis has sharply increased (e.g., the concept of a mean field was used [1-5]). Simultaneously, a number of publications [1-5] concerned analytical study and computer simulation of reactions in different spaces [6-9], including fractal ones [10], has appeared. It has been clarified, that the main factor, defining chemical reactions course, is space connectivity degree, irrespective of its type. Also a large amount of theoretical and applied researches on irreversible aggregation models of different kinds offered for the description of such processes as flocculation, coagulation, and polymerization are carried out [11]. The fractal analysis, intensively developing per last years, as the aggregation within the frameworks of the indicated models forms fractal aggregates [1-11]. Nevertheless, the application of these modern physical concepts for the description of polymers synthesis still has unitary character [12-15].

However, the application of the fractal analysis methods to synthesis process for today becomes a vital problem. Such necessity is not due to the convenience of the fractal analysis as mathematical approach that supposes the existence of approaches, alternate to it. The necessity of the indicated problem solution is defined only by physical reasons. The basic object during synthesis of polymers in solutions is the macromolecular coil, which represents a fractal object [16, 17]. The description of fractal objects within the frameworks of Euclidean geometry is incorrect, that predetermines the necessity of fractal analysis application [18]. Besides, practically all kinetic curves at synthesis of polymers represent curves with the decrease in reaction rate, that is typical designation for fractal reactions [19, 20] and reactions of fractal objects, or reactions in fractal space. Therefore, the purpose of the present review is fractal analysis methods application for synthesis kinetics description and this process final characteristics determination for branched polymers on the example of poly(hydroxyl ether).

2.1.1 The Macromolecular Coil Structure Influence on Poly(hydroxyl ether) Synthesis

The studied poly(hydroxyl ether) (PHE) was synthesized by one step method, namely, by a direct interaction of epichlorohydrin and 4,4`-dioxidiohenylpropane according to the scheme [21]:

The dependences of PHE synthesis were on key characteristics [21], namely, reduced viscosity $\eta_{red.}$ and conversion degree Q, on synthesis temperature T were studied. It was found, that T rising up to the definite limits influences favorably on the indicated process--a reaction rate rises, η_{red} and Q are increased. This effect can be observed in the narrow enough range of $T = 333–348K$. At T lower than 333K PHE formation process decelerates sharply, that is due to insufficient activity of epoxy groups at low temperatures. At $T > 353K$ cross-linking processes proceed, which are due to the activity enhancement of secondary hydroxyls in polymer chain [21]. It is also supposed, that at the indicated temperatures of synthesis PHE branched chains formation is possible [21].

As it is known [16], the macromolecular coil, which is the main structural unit at polymers synthesis in solution, represents a fractal and its structure (coil elements distribution in space) can be described by the fractal dimension D. Proceeding from this, the authors [22] used the fractal analysis methods for the description of T effect on PHE synthesis course and its main characteristics.

The general fractal relationship for synthesis processes description can be written as follows [23]:

$$Q \sim t^{(3-D)/2} \tag{1}$$

where Q is conversion degree, t is synthesis duration.

If the relationship (1) is expressed in a diagram form in double logarithmic coordinates, then from its slope in case of such plot linearity the exponent in the indicated relationship and, hence, the value D, can be determined. The calculated by the indicated mode according to the curves $Q(t)$ for the initial part of these curves values D are adduced in Table 1. As one can see, reduction D at T growth is observed.

TABLE 1 The characteristics of branched PHE [22].

The synthesis temperature, K	D, the equation (1)	D, the equation (15)	d_s	g
333	1.98	1.93	1.29	0.394
338	1.89	1.88	1.24	0.478
343	1.69	1.67	1.03	0.774
348	1.56	1.54	1.0	1.0

Within the frameworks of fractal analysis fractal (macromolecular coil) branching degree is characterized by spectral (fraction) dimension d_s, which is object connectivity degree characteristic [24]. For linear polymer $d_s = 1.0$, for statistically branched one $d_s = 1.33$ [24]. For macromolecular coil with arbitrary branching degree the value d_s varies within the limits of 1.0-1.33. Between dimensions D and d_s the following relationship exists, that takes into consideration the excluded volume effects [17]:

$$D = \frac{d_s(d+2)}{2+d_s},\qquad(2)$$

where d is dimension of Euclidean space, in which a fractal is considered (it is obvious, that in our case $d = 3$).

The Equation (2) allows to estimate the values d_s according to the known magnitudes D (Table 1). As it follows from the data, T increase results to d_s reduction, that is polymer chain branching degree decrease and at $T = 348K$ PHE polymer chain is a linear one ($d_s \approx 1.0$).

A number of traditional estimation methods of polymer chain branching exists as well [25-27]. So, the branching factor g is defined as follows [27]:

$$g = \frac{R_\theta^2}{R_{l,\theta}^2}\qquad(3)$$

where R_θ and $R_{l,\theta}$ are mean gyration radii of a branched polymer and its linear analog in θ solvent at the same values of molecular weight MM and Kuhn segment size A, which characterizes chain thermodynamical rigidity.

Within the frameworks of fractal analysis the relationship between coil gyration radius and molecular weight is given as follows [17]:

$$R_\theta \sim MM^{1/D_\theta},\qquad(4)$$

$$R_{l,\theta} \sim MM^{1/D_{l,\theta}} \tag{5}$$

where D_θ and $D_{l,\theta}$ are macromolecular coil fractal dimensions of branched polymer and its linear analog in θ solvent, respectively.

The fractal equation can be obtained from the relationships (3)-(5) combination for g estimation [22]:

$$g = MM^{2\left[(D_{l,\theta}-D_\theta)/D_{l,\theta}D_\theta\right]} \tag{6}$$

For linear macromolecule the dimension $D_{l,\theta}$ is always equal to 2.0 [16]. For its branched analog D_θ determination is more difficult and this dimension value will depend on the branching degree. For statistically branched coil [28] the following result was obtained:

$$D_\theta = \frac{4(d+1)}{7} \approx 2.286 \tag{7}$$

If we assume, that D_θ changes proportionally to d_s. If boundary conditions, $D_\theta = D_{l,\theta}$ = 2.0 at d_s = 1.0 and D_θ = 2.286 at d_s = 1.33, then according to the obtained above d_s values (Table 1) the corresponding dimension D_θ can be calculated according to the formula [22]:

$$D_\theta = 2 + 0.858(d_s - 1) \tag{8}$$

The values g, calculated according to the Equations (6) and (8), are cited in Table 1. As it was expected, the branching degree growth (g decrease) at T reduction is observed. The cited in Table 1 values g were calculated for $MM = 3 \times 10^4$.

In Figure 1 the dependences $D(g)$ for PHE and also for poly(phenylxalines) (PPX) and bromide-containing aromatic copolyethersulfones (B-PES) were shown [29]. As one can see, for all adduced in Figure 1 polymers D similar growth at g reduction is observed, that indicates on the observed effect community. The different D values for linear analogs at $g = 1.0$ are due to different solvents usage, that is to different level of interactions polymer solvent [22].

The Equation (6) supposes chain branching degree increase (g reduction) at MM growth. In Figure 2 the dependence $g(MM)$, calculated according to the Equation (6), is adduced, which illustrates this law. The calculation g was fulfilled at $D_{l,\theta} = 2.0$ and $D_\theta = 2.249$ [22].

An interactions level between macromolecular coil elements and interactions polymer solvent level can be characterized with the aid of the parameter ε, which is determined as follows [30]:

$$\left\langle \overline{h}^2 \right\rangle \sim MM^{1+\varepsilon} \tag{9}$$

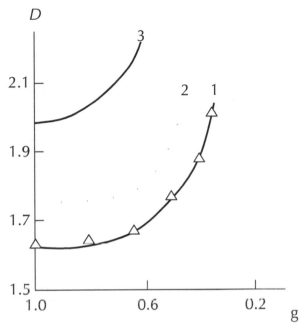

FIGURE 1 The dependences of macromolecular coil fractal dimension D on branching factor g for PHE (1), B-PES (2) and PPX (3) [22].

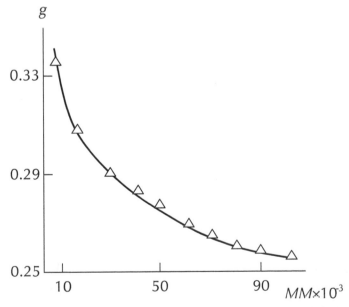

FIGURE 2 The dependence of branching factor g on molecular weight MM for PHE (at $D_{l,\theta} = 2$ and $D_\theta = 2.249$) [22].

where $\langle \bar{h}^2 \rangle$ is mean square distance between macromolecule ends.

The following relationship between D and ε was obtained [31]:

$$D = \frac{2}{\varepsilon + 1}. \tag{10}$$

The ε positive values characterize repulsion forces between coil elements, the negative ones – attraction forces. In Figure 3 the dependence of parameter ε on the branching factor g for PHE is adduced. As it follows from the adduced plot, the dependence $\varepsilon(g)$ is linear and can be described by the following empirical relationship [22]:

$$\varepsilon = 0.45g - 0.15 \tag{11}$$

Therefore, from the Equation (11) it follows, that macromolecule branching degree increase (g decrease) results to attraction forces growth between coil elements and, as consequence, to coil compactization (D growth). Let us note, that the entire variation range ε, corresponding to the same variation range $D = 1$-3 [32], makes up 1/3-1.0 [31]. From the Equation (11) the variation range ε for PHE can be obtained at the condition $g = 0$-1 [25]: $\varepsilon = -0.15$-0.30. This corresponds to variation of $D = 1.54$-2.35 according to the Equation (10). Hence, PHE macromolecule can assume different structural states within the range of leaking coil-coil in θ solvent [16].

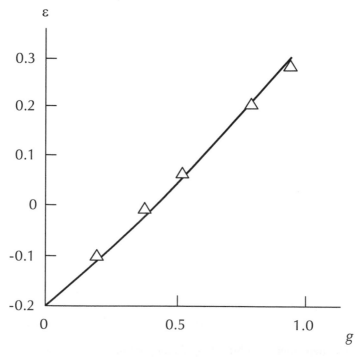

FIGURE 3   The dependence of parameter e on branching parameter g for PHE [22].

The g change results to the exponent a change in Kuhn-Mark-Houwink equation, describing relation between intrinsic viscosity $[\eta]$ and MM [27]:

$$[\eta] = K \cdot MM^a \qquad (12)$$

where K and a are constants for the given polymer.

As it is known [14], the following relationship exists between D and a:

$$D = \frac{3}{1+a} \qquad (13)$$

The Equations (10), (11), and (13) combination allows obtaining the dependence of a on g for PHE [22]:

$$a = 0.275 + 0.68g \qquad (14)$$

Hence, the increase in a chain branching degree (g reduction) results to decrease in a. This conclusion is confirmed experimentally [25] on the example of two polyarylates D-1, received by equilibrium and interphase polycondensation. For the first from them (the branched one) the value a same condition is smaller than for the second (linear one). Besides, for θ conditions in case of linear polyarylate $a = 0.50$, that was expected [16], whereas for branched polyarylate at the same conditions $a = 0.36$. According to the Equation (13) this corresponds to $D_\theta \approx 2.206$, as well adduced above Family estimation [28], according to the Equation (7) and the Equation (8), it corresponds to $d_s \approx 1.24$, that is to a branched polymer.

Using the Equations (10) and (11) combination, the relationship between D and g can be obtained as follows [22]:

$$D = \frac{2}{0.85 + 0.45g} \qquad (15)$$

Calculated according to the Equation (15) D values are also adduced in Table 1, from which their good correspondence to D values, estimated according to kinetic curves $Q(t)$ with the aid of the relationship (1), follows (the mean discrepancy of value D, received by two indicated methods, makes up ~1.4%).

At the end, let us consider, the physical significance of PHE polymer chain branching degree decrease at synthesis temperature growth. The branching centers average number per one macromolecule m can be determined according to the equation [26]:

$$g = \left[\left(1 + \frac{m}{7}\right)^{1/2} + \frac{4m}{9\pi}\right]^{-1/2} \qquad (16)$$

As it is well known [33], fractal objects are characterized by strong screening of internal regions by fractal surface. Therefore, accessible for reaction (in our case for branching formation) sites are either on fractal (macromolecular coil) surface, or near it. Such sites number N_u scales with coil gyration radius R_g as follows [33]:

$$N_u \sim R_g^{d_u} \tag{17}$$

where d_u is dimension of unscreened (accessible for reaction) surface, which is determined according to the equation [33]:

$$d_u = (D-1) + \frac{(d-D)}{d_w} \tag{18}$$

where d_w is dimension of random walk on fractal, which is estimated according to Aarony-Stauffer rule [34]:

$$d_w = D + 1 \tag{19}$$

Besides, it is well known [35], that in case of chemical reactions of various kinds course including reactions at polymers synthesis, the so called steric factor p ($p \leq 1$) plays an essential role, which shows, that not all collisions of reacting molecules occur with proper for chemical bond formation these molecules orientation. The value p is connected with R_g as follows [36]:

$$p \sim \frac{1}{R_g} \tag{20}$$

Therefore, it can be assumed, that the number of accessible for branching formation sites of macromolecular coil m will be proportional to the product pN_u or [22]:

$$[\eta] = \frac{\eta_{red}}{1 + K_\eta c_p \eta_{red}} \tag{21}$$

For this problem solution it is necessary to determine R_g variation at T change. This can be made as follows. Using experimentally determined η_{red} values, $[\eta]$ can be calculated according to Shultze-Braschke empirical equation [37]:

$$[\eta] = \frac{\eta_{red}}{1 + K_\eta c_p \eta_{red}} \tag{22}$$

where K_η is coefficient, which is equal to 0.28 [37] and c_p is polymer concentration.
  Then the coefficient K in Kuhn-Mark-Houwink equation can be determined [25]:

$$K = \frac{21}{m_e} \left(\frac{1}{2500 \cdot m_e} \right)^a \tag{23}$$

where m_e is mean weight of polymer elementary link (without substituent).

Further the value MM is determined according to the Equation (12) and polymerization degree N is calculated as follows:

$$N = \frac{MM}{m_0} \qquad (24)$$

where m_0 is monomer link molecular weight (for PHE $m_0 = 284$).

The last value R_g is calculated according to the following fractal relationship [31]:

$$R_g = 37.5 N^{1/D} \quad \text{Å.} \qquad (25)$$

In Figure 4 the dependence of m on $R_g^{d_u-1}$ is adduced, which prove to be linear. Since the value m increases at $R_g^{d_u-1} \sim pN_u$ growth, then this supposes the offered above treatment identity [22].

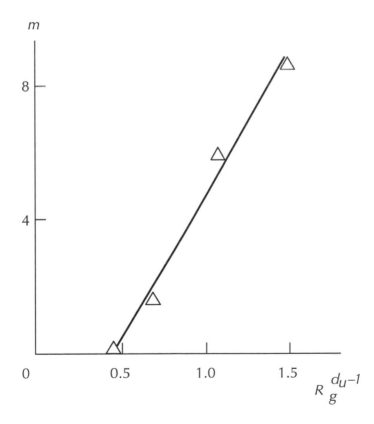

FIGURE 4 The dependence of branching number per one macromolecule m on parameter $R_g^{d_u-1}$ for PHE [22].

Hence, the stated above results showed, that fractal analysis notions allowed to give principally new treatment of phenomena, occurring at PHE synthesis at synthesis temperature variation. The notion of macromolecular coil structure are characterized by its fractal dimension D, forms the basis of this treatment. This structure coil elements interacts among each other and polymer solution, characterized by parameter ε in the relationship (9), are defined. In it is turn, polymer chain branching degree, characterized by branching factor g, is an unequivocal function of D according to the Equation (15). The Equation (2) gives the physical sense of this correlation. Besides, the macromolecular coil structure defines kinetic curves course according to the relationship (1). It is important to note that fractal analysis methods allow the indicated effects quantitative treatment [22]. Proceeding from these general concepts, the authors [38] gave the description of PHE macromolecular coil structure influence on its synthesis rate at four temperatures in the indicated above range of temperatures T.

As it follows from the data of Figure 5, synthesis temperature T increase results to PHE synthesis reaction rate enhancement. Within the frameworks of fractal analysis the reaction rate constant k_r can be determined according to the following relationship [39]:

$$t^{(D-1)/2} = \frac{c_1}{k_r(1-Q)} \qquad (26)$$

where c_1 is constant.

The calculation of polymerization degree N and macromolecular coil gyration radius R_g according to the Equations (24) and (25), respectively, shows wide enough variation of the indicated parameters. The following fractal relationship for steric factor p estimation was obtained [40]:

$$p = \frac{c_2}{t^{(D-1)/2}} \qquad (27)$$

where c_2 is constant, which is also accepted equal to one according to the mentioned above reasons. The values p for PHE at four magnitudes T is given in Table 2.

In Figure 6 the relationship $k_r(p)$ for PHE is adduced, which has an expected character p growth results to k_r increase. However, this dependence is not directly proportional. The authors [38] supposed that for obtaining a more general correlation reaction rate macromolecular coil structure sites number on coil surface N_u, accessible for reaction (unscreened), should be taken into consideration. In other words, in such treatment the dependence of k_r on complex characteristic pN_u should be plotted, which is shown in Figure 7. From this Figure plot it follows, that the linear correlation $k_r(pN_u)$ is now obtained, that is k_r growth at T increase is defined by PHE macromolecular coil structure change.

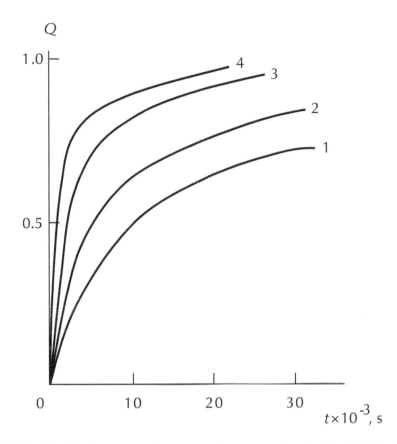

FIGURE 5   The kinetic curves of conversion degree--reaction duration (Q-t) for PHE at synthesis temperatures 333 (1), 338 (2), 343 (3), and 348K (4) [38].

The main studied parameters are given in relative units, then the value c_1 was accepted equal to one [38]. The estimated at $t = 3600$ s k_r values are adduced in Table 2.

TABLE 2   The characteristics of macromolecular coil of PHE, produced at different synthesis temperatures [38].

T, K	k_r, relative units	N	R_g, Å	p
333	0.027	54.4	282	0.0181
338	0.057	97.4	427	0.0262
343	0.205	162.6	764	0.0593
348	0.505	227.4	1215	0.1010

Let us note, that $k_r = 0$ is achieved at small, but finite quantity pN_u. The calculation according to the Equations (18), (26), and (27) shows, that $D \approx 2.20$, that is fractal dimension of branched chain in θ conditions, corresponds to this value pN_u [28]. After this dimension achievement reaction rate decreases sharply.

In Figure 8 the dependence of k_r on the branching factor g, calculated according to the Equation (6) (see Table 1), is adduced in the form of $k_r(g^3)$. As one can see, this dependence is linear and passes through coordinate's origin. Hence, the chain branching degree increase, characterized by g reduction, defines PHE synthesis rate sharp decrease (a cubic dependence).

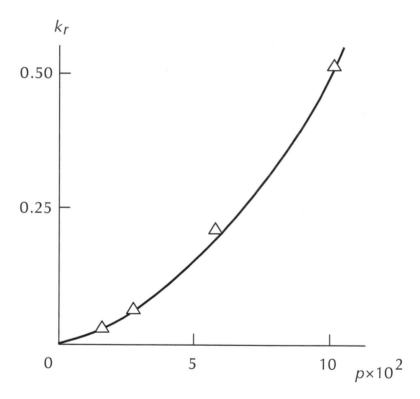

FIGURE 6   The relationship between reaction rate constant k_r and steric factor p for PHE [38].

Hence, the stated results showed definite influence of macromolecular coil structure, characterized by its fractal dimension D, on PHE synthesis rate. The increase in D, that is coil compactization, it decreases sharply reaction rate constant k_r the value in virtue of two factors influence—decrease of accessible for reaction sites number N_u and steric factor p reduction. As a matter of fact, D growth defines transition from PHE synthesis diffusive regime to kinetic one [41]. In its turn, polymer chain branching degree growth rises D and decreases sharply k_r [38].

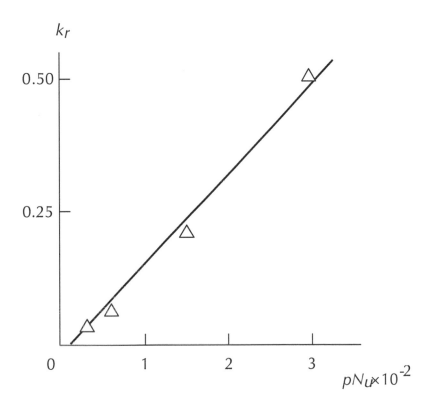

FIGURE 7   The dependence of reaction rate constant k_r on complex characteristic pN_u for PHE [38].

The strong dependence of the reduced viscosity η_{red} on reactionary medium components relation at PHE synthesis in mixture water-isopropanol was found [21]. Besides, it was revealed, that the process of PHE synthesis can be divided into three modes depending on the contents of isopropanol c_{is} in a mixture. At the small contents of isopropanol $c_{is} < 15$ vol.% the synthesis process practically does not take place and it turns into low-molecular oligomer. At $c_{is} = 15$-60 vol.% the linear PHE was received and its viscosity η (or molecular weight MM) grows increasing at c_{is}. And at last, at $c_{is} > 60$ vol.% the cross-linked polymer is formed. The indicated modes existence is explained only from the chemical point of view as follows [21]. The PHE was synthesized by one step method, namely, by direct interaction of epichlorohydrin and bisphenol. The first regime ($c_{is} \leq 15$ vol.%) is explained by poor ability for mixing of epichlorohydrin with water, that complicates its access to bisphenolate, being found in water. Besides, in water the formed products are dropped out as viscous resin and the chain growth is either strongly slowed down or stops.

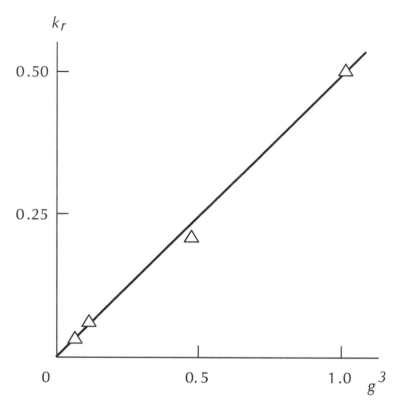

FIGURE 8   The dependence of reaction rate constant kr on branching factor g for PHE [38].

The increase of reactionary ability of hydroxyl group will be promote by OH group introduction in hydrogen bond as the donor of protons owing to the displacement of electronic density under the scheme [42]:

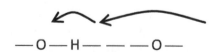

It allows to assume, that at c_{is} = 15-60 vol.% there occurs an increase of bisphenols reactionary ability according to the above mentioned mechanism and this results to the increase of PHE production reaction rate and to higher values η_{red}.

And at last, the production of the cross-linked PHE at c_{is} > 60 vol.% is explained by the participation of isopropanol molecules in reaction, that was observed earlier [43].

The analysis does not take into account the features of macromolecular coil structure, which is a basic element at synthesis of polymers in a solution [40]. The macromolecular coil is a fractal and its structure (the distribution of its links in space) can be characterized by the fractal dimension D. In its turn, the value D is defined by two groups of interactions one by the coil links to each other and another by the polymer solvent [28]. It is evident, that the reactionary medium change should result to the changes of interactions of the second group and, as consequence, to a variation in D. Therefore the authors [44] proposed the alternative explanation of three modes existence during PHE synthesis process with the fractal analysis representations [45, 46] participation.

In Figure 9 the dependence of average value of the reduced viscosity η_{red} for PHE on the isopropanol contents c_{is} in reactionary medium water-isopropanol is adduced and the boundaries of the mentioned synthesis three modes are also indicated (shaded vertical lines). From the data of Figure 9 it follows, that the first regime (synthesis practical absence) is completed at $c_{is} = 15$ vol.%.

The relationship (1) allows to obtain synthesis cessation condition. At $D = d = 3$ (d is the dimension of Euclidean space, in which a process is considered) $Q =$ constant. Besides the initial conditions of synthesis will be $Q = 0$ at $t = 0$. Thus, at $D = d$ $Q =$ constant and the reaction does not take place. In other words, PHE synthesis reaction will be realized in case of fractal reacting objects (macromolecular coils) only.

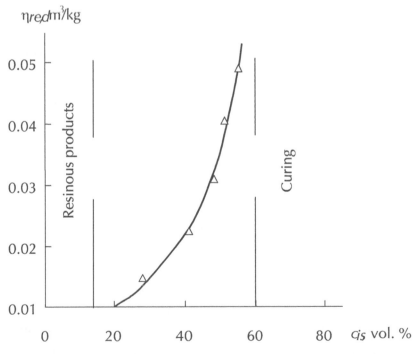

FIGURE 9 The dependence of mean reduced viscosity hred on isopropanol volume contents cis in reactionary mixture water-isopropanol at PHE synthesis. The vertical shaded lines indicate regimes boundary [44].

The value D can be determined according to the equation [47]:

$$D = 1.5 + 0.45\chi_1 \tag{28}$$

where χ_1 is Flory-Haggnis interaction parameter, which characterizes the level of interaction polymer solvent.

In its turn, the value χ_1 is determined according to the following equation [27]:

$$\chi_1 = \frac{E_{ev}}{RT}\left(1 - \frac{\delta_p^2}{\delta_s^2}\right)^2 - \chi_{ent} \tag{29}$$

where E_{ev} is the solvent evaporation heat, R is universal gas constant, T is synthesis temperature, δ_p and δ_s are solubility parameters of polymer and solvent, accordingly, χ_{ent} is the entropic contribution to Flory-Haggins parameter, determined experimentally.

The value δ_s in the considered case is a solubility parameter of mixture water-isopropanol and its determination as a function of c_{is} carried out according to the technique [48]. Within the frameworks of the indicated technique the value δ_s can be written as follows:

$$\delta_s^2 = \delta_f^2 + \delta_c^2, \tag{30}$$

where the solubility parameter component δ_f includes the energy of dispersive interactions and the energy of dipole bonds interaction; the component δ_c includes the energy of hydrogen bonds and the energy of interaction between atom with electrons deficiency of one molecule (acceptor) and atom with electrons abundance of other molecule (donor), which requires the certain orientation of these two molecules. For water, $\delta_f = 8.12$ (cal/cm^3)$^{1/2}$ and $\delta_c = 22.08$ (cal/cm^3)$^{1/2}$, for isopropanol $\delta_f = 7.70$ (cal/cm^3)$^{1/2}$ and $\delta_c = 5.25$ (cal/cm^3)$^{1/2}$ [48].

The value δ_f^m (δ_c^m) for a solvents mixture can be determined according to the mixtures rule [48]:

$$\delta_f^m = \sum_{i=1}^{n} \phi_i \delta_{fi} \tag{31}$$

$$\delta_c^m = \sum_{i=1}^{n} \phi_i \delta_{ci}, \tag{32}$$

where the index "m" designates mixture, the index i designates ith mixture component, the complete number of which is equal to n and ϕ_i is the component volumetric fraction.

The empirical constant χ_{ent} can be estimated according to the following considerations, however the authors [44] showed such estimation necessity. The value D calculation for $c_{is} = 30$ vol.% ($\delta_p = 9.15$ (cal/cm^3)$^{1/2}$, $\delta_s^m = 18.8$ (cal/cm^3)$^{1/2}$) gives the value $D = 3.15$, that is physically impossible, since $D < d$ [49]. This means, that there exists significant entropic contribution to value χ_1 and, hence, D. In Figure 10(a) number of kinetic curves $Q(t)$ for PHE was adduced. Using kinetic curve $Q(t)$ at $T = 348K$ plotting in double logarithmic coordinates, the exponent in the relationship (1) can be estimated and, hence, the value D, which is equal to ~2.60. Then according to the Equation (28), at known E_{ev} (the values E_{ev}^m for mixtures were estimated according to the mixtures rule by analogy with the Equations (31) and (32)), δ_s^m and δ_p can estimate the parameter χ_{ent}, which is equal to ~1.23. In further calculations $\chi_{ent} = $ constant is accepted.

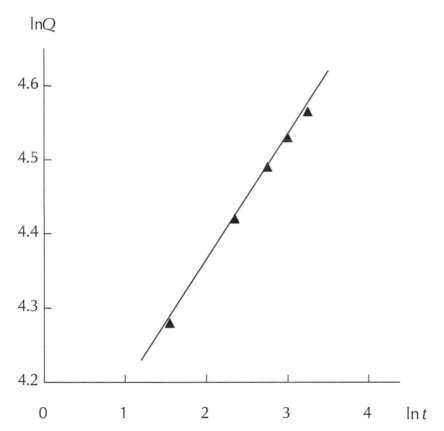

FIGURE 10 The dependence of conversion degree Q on reaction duration t in double logarithmic coordinates for PHE synthesis at T = 348K [44].

In Figure 11 the dependence $D(c_{is})$ calculated by the indicated method, is adduced. At $c_{is} = 15$ vol.% $D = 2.967$, that is macromolecular coil structure, as indicated, does not allow synthesis reaction proceeding. At $c_{is} < 15$ vol.% $D \approx 3 = $ constant in virtue of the mentioned above condition $D \leq d$.

Polymers cross-linking macroscopic process provide joining cluster formation (from one end of a reactionary bath up to another one) [50]. This is gelation process for cross-linking polymer and the cluster dimension at this point is equal to ~2.5 [51, 52]. The gelation process in medium with cross-linked macromolecular coils (so called microgels) are plenty, that results to reactionary medium viscosity essential enhancement and increase in D. According to work [53] the fractal dimension value d_f for such "dense" solution is connected with value D in case of diluted solution by the following relationship:

$$D = \frac{(d+2)d_f}{2(1+d_f)} \qquad (33)$$

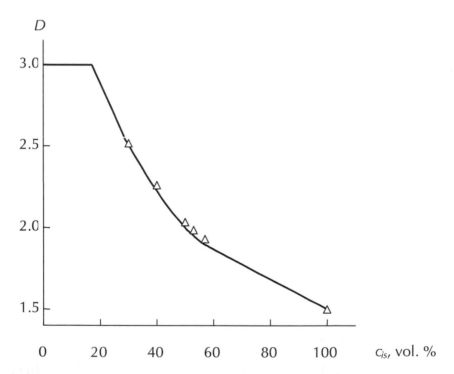

FIGURE 11   The dependence of macromolecular coil fractal dimension D on isopropanol volume contents c_{is} in reactionary mixture water-isopropanol for PHE [44].

For $d_f = 2.5$ we shall receive $D = 1.786$ according to the Equation (33). At $d_f = 3$, that is compact globule, for which any chemical reaction, including cross-linking process, is impossible, we shall receive $D = 1.875$. Hence, the cross-linking process must begin within the range of $D = 1.875$-1.786. According to the Equations (28)-(32) for this value D we shall obtain $\delta_s^m \approx 14.8$ (cal/cm^3)$^{1/2}$, that corresponds to $c_{is} \approx 58$ vol.%. Such estimation corresponds completely to the experimental data, adduced in Figure 9.

Hence, the stated above results showed, that macromolecular coil structure, characterized by its fractal dimension D, could be a critical factor in polymers synthesis regimes definition, in particular, PHE. Compact coil ($D = d$) does not allow reaction proceeding and this condition defines the first regime-resin-like products formation. The decrease in D defines polycondensation reaction proceeding possibility in diluted solutions, and for reaction proceeding on the gelation stage one needs even smaller D value, defining d_f magnitude for "dense" solution. In the considered case the value D is controlled by interactions polymer solvent change.

2.1.2 Theoretical Analysis of Molecular Weight Change in Synthesis Process

As it has been noted, the synthesis temperature increase within the range of $T = 333\text{-}348K$ results to final characteristics enhancement of this process for PHE conversion degree Q and reduced viscosity η_{red}, in the first approximation characterizing polymer molecular weight MM. In Figure 12 the dependences of η_{red} on synthesis duration t are adduced for PHE at four different T. As it follows from these plots, at first η_{red} sharp increase at t growth is observed and then values η_{red} achieve asymptotic branch. Thus, one can suppose, that at some t values η_{red} (or MM) magnitudes achieve their limit, depending on T.

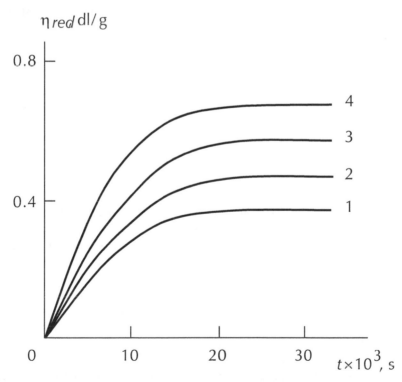

FIGURE 12   The dependences of reduced viscosity hred on synthesis duration t for PHE at synthesis temperature T: 333 (1), 338 (2), 343 (3), and 348K (4) [54].

This effect can be explained within the frameworks of irreversible aggregation models [36], in which a macromolecular coil in solution is considered as a fractal and its structure is characterized by dimension D [16]. The greatest attainable radius R_c of the coil can be estimated according to the following scaling relationship [36]:

$$R_c \sim c_0^{-1/(d-D)} \tag{34}$$

where c_0 is reacting particles (monomers) initial concentration.

The authors [54] estimated applicability of the scaling relationship (34) for the limiting values MM determination and elucidated the macromolecular coil structure, characterized by dimension D, as far as T changes at PHE synthesis.

The range of D values (see Table 1) at PHE synthesis assumes, that this process proceeds according to the mechanism cluster-cluster [36], that is a large macromolecular coil is formed by merging of smaller ones. This circumstance allows to estimate R_c value according to the relationship (34). The value c_0 choice does not influence on the dependence $R_c(D)$ course, but it can change R_c growth rate at D reduction. Proceeding from these considerations [54], the value $c_0 = 25$ was chosen. The experimental values of molecular weight MM^e for PHE can be estimated according to Kuhn-Mark-Houwink Equation (12), which after determination of constants K and a for PHE acquires the following form [54]:

$$[\eta] = 2.84 \times 10^{-4} \left(MM^e \right)^{0.714} \tag{35}$$

Theoretical limiting value of molecular weight MM^T was determined according to the following scaling relationship [17]:

$$MM^T \sim R_c^D \tag{36}$$

In Figure 13 the comparison of the dependences MM^e and MM^T on synthesis temperature T for PHE is adduced. The constant coefficient in the relationship (36) was determined by method of experimental and theoretical values MM superposition. As one can see, a good correspondence of theory and experiment is obtained, that confirms application correctness of the relationship (34) for limiting values MM estimation at PHE synthesis.

Hence, the stated results showed that limiting values of molecular weight, attainable in PHE synthesis process at different T, could be described within the frameworks of irreversible aggregation cluster-cluster model by the usage of the relationship (34). The MM indicated limiting value is controlled by macromolecular coil structure, characterized by its fractal dimension D.

The polymer branching degree influences essentially on molecular weight change kinetics of polymer in its formation process [55, 56]. This process simulation for branched polymers by Monte–Karlo method shows correctness of the following scaling relationship [56]:

$$MM \sim t^{\gamma_t}$$

(37)

where t is reaction duration.

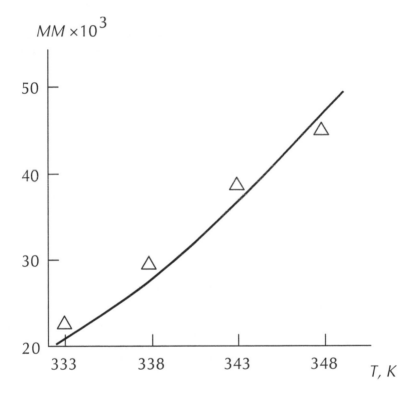

FIGURE 13 The comparison of theoretical (solid line) and experimental (points) dependences of limiting molecular weight MM on synthesis temperature T for PHE [54].

For branched chains, growing in critical conditions, the following relationship was obtained [56]:

$$\gamma_t^{-1} = 1 - g'$$

(38)

where g' is the branching degree, determined as the exponent in the scaling relationship between branching centers number per one macromolecule m and MM [56]:

$$m \sim MM^{g'} \qquad\qquad (39)$$

Hence, the relationship (38) supposes the exponent γ_t growth at polymer branching degree g' increase. The authors [57] obtained the relationship between branching degree and molecular weight in case of real polymer synthesis on the example of PHE.

The experimental values MM for PHE were determined according to the Equation (12) and the constants K and a — according to the formulas (13) and (23), respectively. The values K and a, obtained by the indicated mode, are adduced in Table 3. It is necessary to note, that for the same polymer (PHE) different values of constants in Kuhn–Mark-Houwink were obtained, that is due to different structure of PHE macromolecular coils, received at different T.

TABLE 3 The scaling parameters of PHE, synthesized at different temperatures [57].

T, K	a	K	γ_t^e	γ_t^T
333	0.515	1.4×10^{-3}	0.75	0.606
338	0.587	6.0×10^{-4}	0.54	0.522
343	0.775	7.0×10^{-5}	0.40	0.226
348	0.923	1.3×10^{-5}	0.33	0

Further the dependences $MM(t)$ in double logarithmic coordinates, corresponding to the relationship (37), can be plotted for determination of the exponent γ_t (γ_t^e) experimental values in the indicated relationship. It plotted by the indicated mode dependences $MM(t)$ for four T are shown in Figure 14 and the values γ_t^e are adduced in Table 3. As it follows from these data, the value γ_t^e grows at D increase (see Table 1), that is at polymer chain branching degree enhancement. This situation corresponds completely to conclusions [55, 56]. However, the Equation (38) usage for theoretical value γ_t (γ_t^T) estimation in case of PHE is impossible. Since, $\gamma_t^e < 1$ (see Table 3), then this means negative values g', that does not have physical significance. Therefore, for the estimation of PHE polymer chain branching degree the authors [57] used another parameter–the branching factor g, which is determined according to the Equation (3). Let us note, the principal difference between parameters g' and g, which follows from the relationships (39) and (3) comparison. Polymer branching increase is characterized by g' increase within the range of 0-1 and g decrease within the same range. The values $g' = 0$ and g = 1.0 correspond to linear polymer [25, 27].

Proceeding from the stated above and also from absolute values γ_t^e, the following form of the dependence γ_t^T (g) can be supposed [57]:

$$\gamma_t^T = 1 - g \tag{40}$$

The comparison of γ_t^e and γ_t^T is adduced in Table 3, from which their satisfactory correspondence follows.

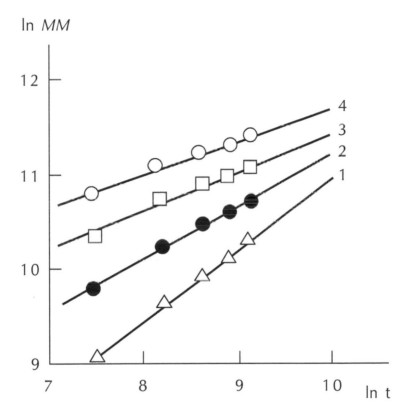

FIGURE 14 The dependences of molecular weight MM on reaction duration t in double logarithmic coordinates for PHE at synthesis temperatures T: 333 (1), 338 (2), 343 (3), and 348K (4) [57].

In Figure 15 comparison of the experimental (calculated according to Kuhn-Mark-Houwink equation) and theoretical (calculated according to the relationship (37) at $\gamma_t = \gamma_t^T$ and for linear PHE at $\gamma_t = \gamma_t^e$) dependences $MM(t)$ is adduced. As one can see, the theory and experiment good correspondence was obtained.

Thus, the stated above results showed correspondence of computer simulation of branched polymers formation and PHE synthesis in the aspect, that the scaling relationship (37) describes correctly molecular weight change kinetics in both cases. However, there exists principal difference in the exponent γ_t determination in the indicated relationship, although in both cases the value γ_t increases at polymer branching degree growth, whichever parameter it is characterized. The indicated discrepancy can be explained by different conditions of branched chains formation.

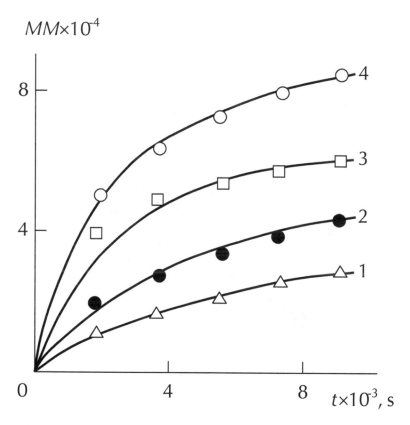

FIGURE 15   The comparison of experimental (lines) and theoretical (points) dependences of molecular weight MM on reaction duration t for PHE at synthesis temperatures T: 333 (1), 338 (2), 343 (3), and 348K (4) [57].

It has been shown, that at PHE synthesis reduced viscosity η_{red} (or polymer molecular weight MM) at reagents initial concentration c_0 growth occurs [21]. Let us note, that similar dependences were also observed at other polymers number synthesis [58]. However, there was no quantitative description of this effect.

The development of lately irreversible aggregation models, elaborated for the description of such processes as polymerization, flocculation, coagulation, and so on, allows to obtain strict physical treatment of similar dependences. Nevertheless, the indicated models application to polymerization real processes are found still rarely enough [59], although it is obvious, that such approach will allow to obtain more profound understanding of aggregation processes, including polymerization.

Therefore, the authors [60] carried out the description of the dependence $\eta_{red}(c_0)$ or $MM(c_0)$ for PHE with the usage of irreversible aggregation models and fractal analysis methods.

In Figure 16 the experimental dependence $\eta_{red}(c_0)$ for PHE (solid line) is adduced, from which monotone growth η_{red} from 0.23 to 0.62 dl/g follows in the indicated range c_0. At $c_0 > 0.7$ mole/l polymer cross-linking process begins, that restricts an upper concentration limit of reaction proceeding [44].

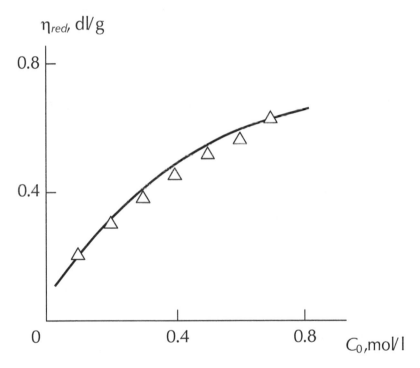

FIGURE 16 The comparison of experimental (solid curve) and theoretical (points) dependences of reduced viscosity hred on initial reagents concentration c0 for PHE [60].

For theoretical description of the dependence $\eta_{red}(c_0)$ a kinetic scaling for diffusion-limited aggregation at the condition $R_g \ll \xi$ (R_g is gyration radius, ξ is a scale of aggregation processes universality classes transition) was used [61]:

$$N \sim \left(c_0 t\right)^{D/(2+D-d)} \tag{41}$$

where N is particles number per aggregate (polymerization degree).

Since, theoretical molecular weight MM^T can be written as $m_0 N$, where m_0 is molecular weight of polymer repeating link (for PHE $m_0 = 284$), then proportionality coefficient in the relationship (41) can be obtained by the curves $MM^e(c_0)$ and $m_0 N(c_0)$

superposition and estimated values MM^T. Then according to these values MM^T using the Equations (35) and (22) the theoretical values η_{red}, corresponding to the model [61], are calculated. In Figure 16 the comparison of the theoretical (points) and experimental (solid line) dependences $\eta_{red}(c_0)$ for PHE is adduced, from which the theory and experiment good correspondence follows.

Hence, the model of irreversible diffusion-limited aggregation allows correct quantitative description of reduced viscosity η_{red} (or molecular weight MM) change with variation of reagents initial concentration c_0. At the condition of value D knowledge and polymer chemical constitution Kuhn-Mark-Houwink equation for it can be received theoretically, that allows to avoid laborious measurements. The relationship (41) predicts that value MM at the fixed t depends not only on c_0, but also on macromolecular coil structure [60].

It has been shown, that characteristic of PHE, synthesized in reactionary medium water-isopropanol, are dependants to a considerable extent on the indicated components ratio [21]. The increase of isopropanol contents in mixture from 30 to 55 vol.% results to reduced viscosity η_{red} growth from ~0.16 to 0.50 dl/g. This effect [21] was explained within the frameworks of Gammet model, which supposed, that isopropyl spirit addition decreased interaction between molecules of water and bisphenol, resulting to enhancement of dioxi compound nucleophyl reactionary ability at the expense of hydrogen bond formation [62].

The adduced above analysis does not take into consideration features of macromolecular coil structure, which is the main element at polymers synthesis in solution. Therefore, the authors [63] explained the considered above effect from the positions of irreversible aggregation models and fractal analysis.

From the relationship (41) it follows, that value of polymerization degree N (and, hence, molecular weight MM) at the fixed c_0 and t is determined only by the fractal dimension D-The smaller D, the larger exponent in the relationship (41) and the higher N (MM). Knowing the value D (see Figure 11) and PHE chemical constitution, Kuhn-Mark-Houwink equation can be received and the experimental values MM (MM^e) according to the experimentally determined η_{red} magnitudes can be calculated. Then, the proportionality coefficient in the relationship (41) can be determined by N and MM^e superposition and thus to obtain theoretical values MM (MM^T), predicted by the model of irreversible aggregation [61]. In Figure 17 the comparison of the dependences $MM^e(c_{is})$ and $MM^T(c_{is})$ for PHE is adduced (in case of MM^e the error limits are adduced [21]), from which theory and experiment good correspondence follows. This indicates to PHE synthesis process description equivalency within the frameworks of irreversible aggregation models.

Since PHE, produced in different mixtures water-isopropanol, has different values D (i.e., different structure), then from the Equations (13) and (23) it follows, that for them values a and K will be different, that is the relation between [η] and MM for the same polymer will be defined by different Kuhn–Mark-Houwink equations. Therefore, strictly speaking, polymer viscosity should be determined in the same solvent, in which it was synthesized. This rule is confirmed by a well known fact [25], that synthesized by different polycondensation modes polyarylates, having the same chemical constitution, have different values a (and, hence, D according to the Equation (13)) and K and also distinguishing properties. In Figure 17 the dependence $MM^e(c_{is})$, calculated according to the same Kuhn-Mark-Houwink, is adduced. As it follows from the data of Figure 17, MM calculation accounting for coil structure and without it gives close enough results at large MM, but at small MM ($<10^4$) the discrepancy can be even quintuple one.

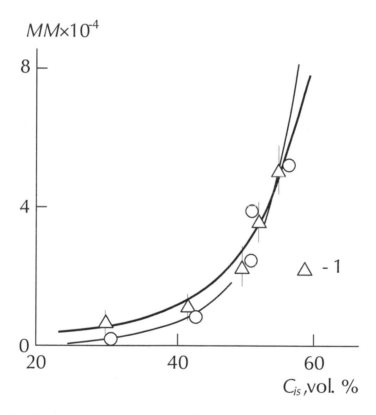

FIGURE 17   The comparison of experimental (1) and calculated according to the relationship (41) at using of different (2) and one (3) Kuhn-Mark-Houwink equations dependences of molecular weight MM on isopropanol contents cis in reactionary mixture for PHE synthesis [63].

Hence, the stated above results showed that PHE molecular weight changes at water-isopropanol ratio variation in reactionary mixture could be described quantitatively within the frameworks of irreversible aggregation models and fractal analysis. The indicated ratio change defines interactions polymer solvent change that results to macromolecular coil structure variations. An interactions polymer solvent weakness, characterized by Flory-Haggins interaction parameter χ_1 reduction, results to

D decreasing, that makes macromolecular coil more accessible for synthesis reaction proceeding.

2.1.3   The Branching Degree and Macromolecular Coil Structure

As it has been noted above, the branching degree of polymer chain can be characterized by several parameters. One of them is a number of branching centers per one macromolecule m. The branching degree g', determined from the scaling relationship (39), serves as another parameter. As a rule, the value $g' < 1$ [27] and this means that the number of branching centers is not proportional to the length of macromolecule or its polymerization degree, N. From the chemical point of view, such effect is difficult to explain, since each monomer link in macromolecule has the same probability of branch formation, and then one can expect $m \sim N$. However, in the real conditions of polymers synthesis there are a number of causes which can in principle cause the ratio m/N to decrease. One of such reasons can be the fact that the branching reactive centers, formed in the initial stages of synthesis, are proved to be "buried" inside a macromolecular coil and, consequently, are less accessible [55]. Such situation defines the necessity of the macromolecular coil structure allowance that can be fulfilled with the aid of its fractal dimension D. Therefore, the authors [64] carried out the description of the macromolecular coil structure influence on accessible for reaction branching centers number at polymer molecular weight change. This description is given within the frameworks of fractal analysis on the example of PHE.

Besides, the characteristics indicated above one more parameter, the branching factor g, can be used for estimation of polymer branching degree, which is determined according to the Equation (3). Within the frameworks of fractal analysis the formula (6) allows to determine the value g, which supposes the dependence of g on molecular weight. The parameters g and m are connected by the relationship (16).

In Figure 18 the dependence $m(MM)$, where the value m was calculated according to the Equations (6) and (16), is adduced. As one can see, this dependence is a nonlinear one that is the value m grows much weaker than MM. In Figure 19 the same dependence is adduced in double logarithmic coordinates, which proves to be linear, and from its slope the value $g' \approx 0.272$ can be estimated. The small value g' supposes strong influence of macromolecular coil structure on m value and this effect can be estimated quantitatively as follows. One of the main features of the fractal object structure is strong screening of its internal regions by the surface [33]. Therefore, the accessible for chain branching reaction macromolecular coil sites are disposed either on its surface or near it. The number of such sites N_u is determined according to the scaling relationship (17).

The dependence $m(R_g^{d_u-1})$, corresponding to the relationship (21), is shown in Figure 4. As one can see, it is linear and has an expected character: m increases with $R_g^{d_u-1}$ enhancement. The extrapolation of this plot to $m = 0$ gives $R_g^{d_u-1} \approx 2.2$, which corresponds to the smallest size of PHE macromolecule for the beginning of the branching. This size is equal to ~9.9 Å. The volume of the repeating link of PHE V_0 can be estimated according to the equation [65]:

$$V_0 = \frac{m_0}{\rho N_A} \qquad (42)$$

where ρ is the polymer density (for PHE, $\rho \approx 1150$ kg/m^3 [6]) and N_A is Avogadro's number.

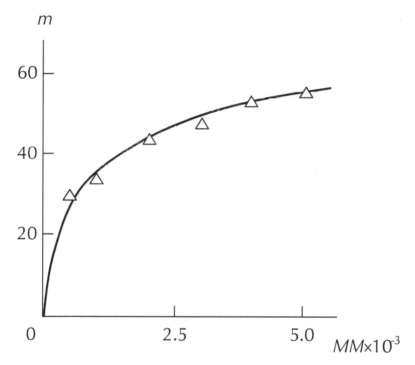

FIGURE 18 The dependence of branching center number per one macromolecule m on molecular weight MM for PHE synthesized at T = 333K [64].

Then, believing that the cross sectional area of PHE macromolecule is equal to 30.7 Å^2 [67] and using the volume $V_0 \approx 410$ Å^3, calculated according to the Equation (42), the monomer link length for PHE, l_0, can be estimated to be equal to ~13.4 Å. The comparison of the smallest value R_g and l_0 indicates that PHE chain branching process begins already at the initial synthesis stage.

Let us note, that the dependences $m(MM)$ with fractional exponent $g < 1$ are typical for other polymers as well. So, in work [25] the constants of Kuhn-Mark-Houwink equation for θ solvent in case of the branched polyarylate D-1 and its linear analog have been reported, that allows to calculate intrinsic viscosities $[\eta]_\theta$ and $[\eta]_{l,\theta}$, accordingly, for arbitrary *MM*. Then, the value of *g* can be estimated according to the relationship [25]:

$$\frac{[\eta]_\theta}{[\eta]_{l,\theta}} = g^{2-a}$$ (43)

where *a* is an exponent in Kuhn-Mark-Houwink equation for a linear analog in θ solvent (*a* = 0.5 [25]).

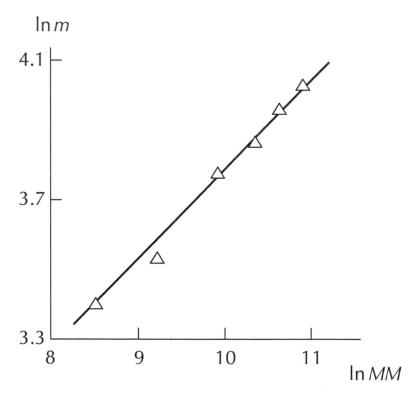

FIGURE 19 The dependence of branching center number per one macromolecule m on molecular weight MM in double logarithmic coordinates for PHE synthesized at T = 333K [64].

Knowing the values *g*, *m* magnitudes can be calculated according to the Equation (16). The estimations have shown, that for polyarylate *D*-1 at *MM* increase from 5×10^4 up to 10×10^4 *m* growth from 2.0 to 2.8 is observed, which corresponds to g ≈ 0.66 [64].

Taking into consideration that gyration radius, R_g, scales to *MM* according to the Equations (4) and (5), from the relationship (21) one obtains [64]:

$$m \sim MM^{(d_u-1)/D} \qquad (44)$$

The comparison of the relationships (39) and (44) allows receiving the following equation [64]:

$$g = c_{ch}\left(\frac{d_u-1}{D}\right) \qquad (45)$$

The proportionality coefficient, c_{ch}, in the Equation (45) has a clear physical significance: it defines the greatest density of "chemical" branching centers per one macromolecule. Parameter $(d_u-1)/D$ define this density decrease by macromolecular coil structural features. For PHE the value $c_{ch} \approx 1.41$. The values of g, $(d_u-1)/D$ and $c_{ch}[(d_u-1)/D]$ for PHE at three synthesis temperatures T are listed in Table 4, from which a good correspondence of the first and the third parameters from the indicated ones follows.

TABLE 4   The experimental and theoretical characteristics of PHE chain branching at different synthesis temperatures [64].

T, K	G	$\dfrac{d_u-1}{D}$	$c_{ch}(\dfrac{d_u-1}{D})$
333	0.272	0.193	0.270
338	0.229	0.145	0.203
343	0.130	0.105	0.147

For branched polyarylate D-1, D value can be determined according to the Equation (13). Further, according to the Equations (18) and (19), by using $a = 0.36$ [25] and $D \approx 2.20$, d_u can be determined and c_{ch} can be calculated for D-1 according to the Equation (45). In this case $c_{ch} \approx 3.22$, that is the greatest "chemical" branching centers density for D-1 is much higher than for PHE.

At the end, the integral dependence of m on chemical and physical factors can be written [64]:

$$m \sim MM^{c_{ch}[(d_u-1)/D]} \qquad (46)$$

The dependence corresponding to the relationship (46) is shown in Figure 20. As one can see, now this correlation is linear and passes through the coordinate's origin. This allows to assert that all factors controlling m value are taken into consideration [64].

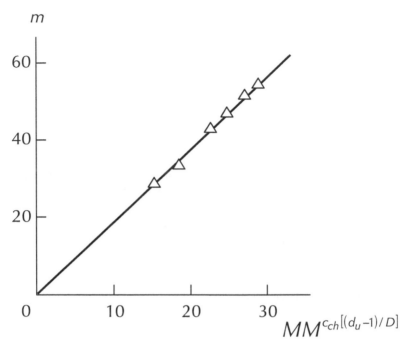

FIGURE 20   The dependence of branching center number per one macromolecule m on complex parameter $MM^{c_{ch}[(d_u-1)/D]}$ for PHE synthesized at T = 333K [64].

Hence, the fractal analysis methods are efficient for clear structural identification of both chemical and physical factors, controlling a chain branching degree. The number of effective branching centers per one macromolecule m is controlled by four factors: polymer molecular weight MM, maximum "chemical" density of reactive centers c_{ch}, dimension of unscreened surface d_u of macromolecular coil, and its fractal dimension D. The Equation (45) allows to determine the critical value $D(D_{cr})$, below of which $g = 0$ (i.e., branching does not occur): $D_{cr} = 1.10$ [64].

KEYWORDS

- **Aarony-Stauffer rule**
- **Avogadro's number**
- **Flory-Haggins parameter**
- **Isopropanol**
- **Kuhn-Mark-Houwink equation**
- **Macromolecular coil**
- **Polyphenylxalines**

REFERENCES

1. Belyi, A. A. and Ovchinnikov, A. A. Report of Academy of Sciences of USSR (in rus.), **288** (1), 151-155 (1986).
2. Burlatskii, S. F. Report of Academy of Sciences of USSR (in rus.), **288** (1), 155-159 (1986).
3. Aleksandrov, I. V. and Pazhitnov, A. V. *J. Chem. Phys.* (in rus.), 6(9), 1243-1247 (1987).
4. Burlatskii, S. F., Ovchinnikov, A. A., and Pronin, K. A. *J. Exp. and Theo. Phys.* (in rus.), 92(2), 625-637 (1987).
5. Burlatskii, S. F., Oshanin, G. S., and Likhachev, V. N. *J. Chem. Phys.* (in rus.), 7(7), 970-978 (1988).
6. Grassberger, P. and Procaccia, I. *J. Chem. Phys.*, 77(12), 6281-6284 (1982).
7. Havlin, S., Weiss, G. H., Kiefer, J. E., and Dishon, M. *J. Phys. A*, 17(4), L347-350 (1984).
8. Redner, S. and Kang, K. *J. Phys. A*, 17(5), L451-455 (1984).
9. Kang, K. and Redner, S. *Phys. Rev. Lett.*, 52(12), p. 955-958 (1984).
10. Meakin, P. and Stanley, H. E. *J. Phys. A*, 17(2), L173-177 (1984).
11. Shogenov, V. N. and Kozlov, G. V. *The Fractal Clusters in Physics-Chemistry of Polymers.* (in rus.) Polygraphservice Publishing House, Nal'chik-city, p. 268 (2002).
12. Kaufman, J. H., Baker, C. K., Nazzal, A. I., Flickner, M., Melray, O. R., and Kapitulnik, A. *Phys. Rev. Lett.*, 56(18), 1932-1935 (1986).
13. Chu, B., W. C., Wu, D. Q., and Phillips, J. C. *Macromolecules*, 20(10), 2642-2644 (1987).
14. Karmanov, A. P. and Monakov, Yu. B. *Russian Polymer Science (in rus.).* B, 37(2), 328-331 (1995).
15. Kozlov, G. V., Shustov, G. B., and Zaikov, G. E. *J. Appl. Polymer Sci.*, 111(7), 3026-3030 (2009).
16. Baranov, V. G., Frenkel, S. Ya., and Brestkin, Yu. V. Report of Academy of Sciences of USSR (in rus.), **290**(2), 369-372 (1986).
17. Vilgis, T. A. *Phys. Rev. A*, 36(3), 1506-1508 (1987).
18. Rammal, R. and Toulouse, G. *J. Phys. Lett. (Paris)*, 44(1), L13-22 (1983).
19. Klymko, P. W. and Kopelman, R. *J. Phys. Chem.*, 87 (23), 4565-4567 (1983).
20. Kopelman, R., Klymko, P. W., Newhouse, J. S., and Anacker, L. W. *Phys. Rev. B*, 29(6), 3747-3748 (1984).
21. *Monomers, Oligomers, Polymers, Composites and Nanocomposites Research. Synthesis, Properties and Applications*, Ed. by R.A. Pethrick, P. Petkov, G.E. Zaikov, S.K. Rakovsky, Polymer Yearbook, Vol. 23, 2012, 481 pp., second edition.
22. Kozlov, G. V., Beeva, D. A., and Mikitaev, A. K. News in Polymers and Polymer Composites (in rus.), **1**(1), (2011), p. 3.
23. Novikov, V. U. and Kozlov, G. V. *Russian Chemical Review* (in rus.), 69(4), 378-399 (2000).
24. Alexander, S. and Orbach, R. *J. Phys. Lett.* (Paris), 43(17), L625-631 (1982).
25. Askadskii, A. A. *The Physics-Chemistry of Polyarylates.* Chemistry Publishing House (in rus.), Moscow, p. 216 (1968).
26. Korshak, V. V., Pavlova, S. S. A., Timofeeva, G. I., Kroyan, S. A., Krongauz, E. S., Travnikova, A. P., Raubah, H., Shultz, G., and Gnauk, R. *Russian Polymer Science (in rus.),.* A, 24(9), 1868-1876 (1984).
27. Budtov, V. P. *Physical Chemistry of Polymer Solutions(in rus.).* Chemistry Publishing House, St-Peterburg, p. 384 (1992).
28. Family, F. *J. Stat. Phys.*, 36(5/6), 881-896 (1984).
29. Kozlov, G. V., Mikitaev, A. K., and Zaikov, G. E. *Polymer Research J.*, 2(4), 381-388 (2008).
30. Pavlov, G. M. and Korneeva, E. V. *Biophysical Journal (in rus.)*, 40(6), 1227-1233 (1995).
31. *Polymer Yearbook – 2011. Polymers, Composites and Nanocomposites. Yesterday, Today, Perspectives*, Ed. by G.E. Zaikov, C. Sirghie, R.M. Kozlowski, Nova Science Publishers, New York, 2012, 254 pp.
32. Kozlov, G. V., Afaunova, Z. I., and Zaikov, G. E. *Polymer International*, 54(4), 1275-1279 (2005).

33. Meakin, P., Coniglio, A., Stanley, E. H., and Witten, T. A. *Phys. Rev. A*, **34**(4), 3325-3340 (1986).
34. Sahimi, M., McKarnin, M., Nordahl, T., and Tirrell, M. *Phys. Rev. A*, **32**(1), 590-595 (1985).
35. Barns, F. S. *Biophysical Journal (in rus.)*, **41**(4), 790-802 (1996).
36. Kokorevich, A. G., Gravitis, Ya. A., and Ozol-Kalnin, V. G. *Chem. W. J.* (in rus.), (1), 3-24 (1989).
37. Brown, D., Sherdron, G., and Kern, V. *Practical Hand book by Synthesis and Study of Polymers Properties*. V. P. Zubkov (Ed.). Chemistry Publishing House (in rus.), Moscow, p. 256 (1976).
38. G.E. Zaikov, R.M. Kozlowski *Chemical reactions in gas, liquid and solid phases: synthesis, properties and application*, Nova Science Publishers, New York, 2012, 282
39. Kozlov, G. V., Shustov, G. B., and Zaikov, G. E. *J. Appl. Polymer Sci.*, **935**, 2343-2347 (2004).
40. Kozlov, G. V. and Shustov, G. B. In *The Achievements in Polymers Physics-Chemistry Field*, G. E. Zaikov (Ed.). Chemistry Publishing House (in rus.), Moscow, pp. 341-411 (2004).
41. Jullien, R. and Kolb, M. *J. Phys. A*, **17**(12), L639-643 (1984).
42. Sorokin, M. F. and Chebotareva, M. A. *Pl. M. J. (in rus.)*, (5), 8-10 (1985).
43. Shvets, V. F. and Lebedev, N. N. *Proceedings of Moscow Institute of Fine Chemical Technology (in rus.)*, (42), 72-77 (1963).
44. Kozlov, G. V. and Zaikov, G. E. *J. Balkan Tribologic. Assoc.*, **9**(2), 196-202 (2003).
45. Kozlov, G. V. and Zaikov, G. E. *Fractal Analysis and Synergetics of Catalysis in Nanosystems*. Nova Publishers, New York, p. 163 (2008).
46. Naphadzokova, L. Kh. and Kozlov, G. V. *Fractal Analysis and Synergetics of Catalysis in Nano-systems*. Nauka (Science) Publishing House (in rus.), Moscow, p. 230 (2009).
47. Kozlov, G. V., Shustov, G. B., and Dolbin, I. V. Proceedings I Internat. Sci. Conf. *Modern Problems of Organic Chemistry, Ecology and Biotechnology*. Nauka (Science) Publishing House (in rus.), Luga-town, 17-18 (2001).
48. Wiehe, I. A. *Ind. Engng. Chem. Res.*, **32**(2), 661-673 (1995).
49. Mandelbrot, B. B. *The Fractal Geometry of Nature*. W. H. Freeman and Company, San-Francisco, p. 459 (1982).
50. Hess, W., Vilgis, T. A., and Winter, H. H. *Macromolecules*, **21**(8), 2536-2542 (1988).
51. Botet, R., Jullien, R., and Kolb, M. *Phys. Rev. A*, **30**(4), 2150-2152 (1984).
52. Kobayashi, M., Yoshioka, T., Imai, M., and Itoh, Y. *Macromolecules*, **28**(22), 7376-7385 (1995).
53. Muthukumar, M. *J. Chem. Phys.*, **83**(6), 3161-3168 (1985).
54. Kozlov, G. V., Grineva, L. G., and Mikitaev, A. K. *Mater of VI Internat. Sci. Pract. Conf. New Polymer Composite Materials*. Kabardino-Balkarian State University Publishing House (in rus.), Nal'chik-city, p. 200-204 (2010).
55. Alexandrowich, Z. *Phys. Rev. Lett.*, **54**(13), 1420-1423 (1985).
56. Alexandrowich, Z. In *Fractals in Physics*. L. Pietronero and E. Tosatti (Ed.). Amsterdam, Oxford, New York, Tokyo, North-Holland, pp. 172-178 (1986).
57. Kozlov, G. V., Shustov, G. B., Mikitaev, A. K. *Mater of V Internat. Sci. Pract. Conf. New Polymer Composite Materials*. Kabardino-Balkarian State University Publishing House (in rus.), Nal'chik-city, 2009, p. 117-123.
58. Korshak, V. V. and Vinogradova, S. V. *A Nonequilibrium Polycondensation*. Nauka (Science) Publishing House (in rus.), Moscow, p. 696 (1972).
59. Magomedov, G. M., Kozlov, G. V., and Zaikov, G. E. *Structure and Properties of Cross-linked Polymers*. A Smithers Group Company, Shawbury, p. 492 (2011).
60. Kozlov, G. V. and Mikitaev, A. K. *Mater of VI Internat. Sci. Pract. Conf. New Polymer Composite Materials*. Kabardino-Balkarian State University Publishing House (in rus.), Nal'chik-city, p. 205-210 (2010).
61. Henschel, H. G. E., Deutch, J. M., and Meakin, P. *J. Chem. Phys.*, **81**(5), 2496-2502 (1984).
62. Gammet, L. *The Principles of Physical Organic Chemistry*. Mir Publishing House (in rus.), Moscow, p. 326 (1972).
63. Kozlov, G. V. and Mikitaev, A. K. *Mater of VI Internat. Sci. Pract. Conf. New Polymer Composite Materials*. Kabardino-Balkarian State University Publishing House (in rus.), Nal'chik-city, p. 192-199 (2010).

64. Kozlov, G. V., Burya, A. I., and Shustov, G. B. *Chemical Industry and Chemical Engineering Quarterly*, **14**(3), 181-184 (2008).
65. Kozlov, G. V. and Sanditov, D. S. *Anharmonic Effects and Physical-Mechanical Properties of Polymers*. Nauka (Science) Publishing House (in rus.), Novosibirsk, p. 261 (1994).
66. Beeva, D. A., Mikitaev, A. K., Zaikov, G. E., and Beev, A. A. *In Molecular and High Molecular Chemistry*, Yu. B. Monakov and G. E Zaikov (Ed.). Nova Science Publishers, Inc., New York, pp. 49-54 (2006).
67. Aharoni, S. M. *Macromolecules*, **18**(12), 2624-2630 (1985).

3 Nanotechnologies in Polymer Processing

A. A. Panov, G. E. Zaikov, and A. K. Panov

CONTENTS

3.1 INTRODUCTION

Composite materials, nanotechnology, prefix nano, superparamagnetizm, thermo-plastic matrix Technical and technological development demands the creation of new materials, which are stronger, more reliable and more durable, that is materials with new properties. Up-to-date projects in creation of new materials go along the way of nanotechnology.

Nanotechnology can be referred to as a qualitatively new round in human progress. This is a wide enough concept, which can concern any area–Information technologies, medicine, military equipment, robotics, and so on. We narrow the concept of nanotechnology and consider it with the reference to polymeric materials as well as composites on their basis.

The prefix nano means that in the context of these concepts the technologies based on the materials, elements of constructions and objects are considered, whose sizes make 10^{-9} meters. As fantastically, as it may sound but the science reached the nanolevel long ago. Unfortunately, everything connected with such developments and technologies for now is impossible to apply to mass production because of their low productivity and high cost.

That means nanotechnology and nanomaterials are only accessible in research laboratories for now but it is only a matter of time. What sort of benefits and advantages will the manufacturers have after the implementation of nanotechnologies in their manufactures and starting to use nanomaterials? Nanoparticles of any material have absolutely different properties rather than micro or macroparticles. This results from the fact that alongside with the reduction of particles' sizes of the materials to nanometric sizes, physical properties of a substance change too. For example, the transition of palladium to nanocristalls leads to the increase in its thermal capacity in more than 1,5 times; it causes the increase of solubility of bismuth in copper in 4,000 times and the increase of self-diffusion coefficient of copper at a room temperature on 21 orders.

Such changes in properties of substances are explained by the quantitative change of surface and volume atoms' ratio of individual particles that is by the high-surface area. Insertion of such nanoparticles in a polymeric matrix while using the apparently old and known materials gives a chance of receiving the qualitatively and quantitatively new possibilities in their use.

Nanocomposites based on thermoplastic matrix and containing natural, laminated inorganic structures are referred to as laminated nanocomposites. Such materials are produced on the basis of ceramics and polymers however with the use of natural laminated inorganic structures such as, montmorillonite or vermiculite which are present for example, in clays. A layer of filler ~1 nm thick is saturated with monomer solution and later polymerized. The laminated nanocomposites in comparison with initial polymeric matrix possess much smaller permeability for liquids and gases. These properties allow applying them to medical and food processing industry. Such materials can be used in manufacture of pipes and containers for the carbonated beverages.

These composite materials are eco-friendly, absolutely harmless to the person and possess fire-resistant properties. The derived thermoplastic laboratory samples have been tested and really confirmed those statements.

Manufacturing technique of thermoplastic materials causes difficulties for today, notably dispersion of silicate nanoparticles in monomer solution. To solve this problem it is necessary to develop the dispersion technique, which could be transferred from laboratory conditions into the industrial ones.

What advantages the manufacturers can have, if they decide to reorganize their manufacture for the use of such materials, can be predicted even today. As these materials possess more mechanical and gas barrier potential in comparison with the initial thermoplastic materials, then their application in manufacture of plastic containers or pipes will lead to raw materials' saving by means of reduction of product thickness.

The improvement of physical and mechanical properties allows applying nanocomposite products under higher pressures and temperatures. For example, the problem of thermal treatment of plastic containers can be solved. Another example of the application of the valuable properties of laminated nanocomposites concerns the motor industry. American company General Motors manufactures the laminated nanocomposite material for the production of parts for the car body of Hummer H$_2$.

Ultrathin nanoclay surpasses talc and other fillers in its quality: it forms particles, which are more rigid and less fragile at low temperatures and approximately for 20% easier. The application of such material allows lowering the weight of the car and raising the durability of its parts.

Another group of materials is metal containing nanocomposites. Thanks to the ability of metal particles to create the ordered structures (clusters), metal containing nanocomposites can possess a complex of valuable properties. The typical sizes of metal clusters from 1 to 10 nm correspond to their huge specific surface area. Such nanocomposites demonstrate the superparamagnetizm and catalytic properties; therefore, they can be used while manufacturing semiconductors, catalysts, optical, and luminescent devices, and so on.

Such valuable materials can be produced in several ways, for example, by means of chemical or electrochemical reactions of isolation of metal particles from solutions. In this case, the major problem is not so much the problem of metal restoration but the preservation of its particle, that is the prevention of agglutination and formation of large metal pieces.

For example under laboratory conditions metal is deposited in such a way on the thin polymeric films capable to catch nanosized particles. Metal can be evaporated by means of high energy and nanosized particles can be produced, which then should be preserved. Metal can be evaporated while using explosive energy, high-voltage electric discharge, or simply high temperatures in special furnaces.

The practical application of metal-containing nanocomposites (not going into details about high technologies) can involve the creation of polymers possessing some valuable properties of metals. For example, the polyethylene plate with the tenth fractions of palladium possesses the catalytical properties similar to the plate made of pure palladium.

An example of applying metallic composite is the production of packing materials containing silver and possessing bactericidal properties. By the way, some countries have already been applying the paints and the polymeric coverings with silver nanoparticles. Owing to their bactericidal properties, they are applied in public facilities (painting of walls, coating of handrails etc.).

At present, independent works on creation of concentrates containing stable silver nanoparticles are carried out parallel in Russia and the Ukraine. On addition of such concentrates to plastics, the latter obtain bactericidal properties therewith opening new prospects in the field of packing materials and water purification. In the near future, such concentrates can become accessible to mass manufacturers of products from plastics.

The technology of polymeric nanocomposites manufacture forges ahead; its development is directed to simplification and cheapening the production processes of composite materials with nanoparticles in their structure. For example, Hybrid Plastics/USA has developed POSS additives (polyhedral oligomeric silsesquioxanes mineral fillers based on silicon) in the form of crystalline solids, liquids, and oils. More recently, these products were estimated in a range of 1,000 dollars per a pound (0.454 kg) however now they cost already 50 dollars per a pound. It is to note that nanotechnology is still a very young scientific field and is at the stage of primary development, as the basic part of information and knowledge is applied only in laboratory conditions. However, the nanotechnologies develop high rates; what seemed impossible yesterday, will be accessible to the introduction on a commercial scale tomorrow [1–14].

The prospects in the field of polymeric composite materials upgrading are retained by nanotechnologies. Ever increasing manufacturers' demand for new and superior materials stimulates the scientists to find new ways of solving tasks on the qualitatively new nanolevel.

The desired event of fast implementation of nanomaterials in mass production depends on the efficiency of cooperation between the scientists and the manufacturers in many respects. Today's high technology problems of applied character are successfully solved in close consolidation of scientific and business worlds.

KEYWORDS

- **Composite materials**
- **Nanotechnology**
- **Prefix nano**
- **Superparamagnetizm**
- **Thermoplastic matrix**

REFERENCES

1. Zaikov, G. E. and Kozlowski, R. M. *Chemical reactions in gas, liquid and solid phases: synthesis, properties and application*. Nova Science Publishers, New York, p. 282 (2012).
2. Stoyanov, O. V., Kubica, S., and Zaikov, G. E. *Polymer material science and nanochemistry*. Institute for Engineering of Polymer Materials and Dyes Publishing House, Torun (Poland) (2012).
3. *The problems of nanochemistry for the creation of new materials*. By A. M. Lipanov, V. I. Kodolov, S. Kubica, and G. E. Zaikov (Eds.), Institute for Engineering of Polymer Materials and Dyes Publishing House, Torun, Poland, p. 252 (2012).
4. *Unique Properties of Polymers and Composites: Pure and Applied Science Today and Tomorrow*. By Yurii N. Bubnov, Valerii A. Vasnev, Andrei A. Askadskii, Gennady E. Zaikov (Eds.) Vol. 1, Nova Science Publishers, New York, p. 278 (2012).
5. *Unique Properties of Polymers and Composites: Pure and Applied Science Today and Tomorrow*. By Yurii N. Bubnov, Valerii A. Vasnev, Andrei A. Askadskii, Gennady E. Zaikov (Eds.), Vol. 2, New York, (2012).
6. *Kinetics, catalysis and mechanism of chemical reactions: From pure to applied science. Volume 1. Today and Tomorrow*. By Regina M. Islamova, Sergey V. Kolesov, and Gennady E. (Eds.), Nova Science Publishers, New York, p. 312 (2012).
7. *Kinetics, catalysis and mechanism of chemical reactions: From pure to applied science. Volume 1. Today and Tomorrow*. By Regina M. Islamova, Sergey V. Kolesov, and Gennady E. (Eds.), Nova Science Publishers, New York, p. 444 (2012).
8. *Modern problems in biochemical physics: New horizons*. By Sergei D. Varfolomeev, Elena B. Burlakova, Anatoly A. Popov, and Gennady E. Zaikov (Eds.), Nova Science Publishers, New York, p. 330 (2012).
9. *Biochemical physics and biodeterioration: New horizons*. By Stefan Kubica, Gennady E. Zaikov, LinShu Liu (Eds.), Institute for Engineering of Polymer Materials and Dyes Publishing House, Poland, p.292 (2012).
10. *Biodamages and their sources for some materials*. By C. Kubica, G. E. Zaikov, E. L. Pekhtasheva (Eds.), Torun, Institute for Engineering of Polymer Materials and Dyes Publishing House, Poland, p. 248 (2012).
11. *Biochemistry and biotechnology: Research and development*. By S. D. Varfolomeev, G. E. Zaikov, L. P. Krylova (Eds.), Nova Science Publishers, New York, p. 352 (2012).

12. *Polymers, Composites and Nanocomposites. Synthesis, Properties and Applications.* By S. Kubica, O. V. Stoyanov, G. E. Zaikov (Eds.), Institute for Engineering of Polymer Materials and Dyes, Torun, p. 202 (2012).
13. *Electrospinning of Polymeric Nanofibers: The Researchers Oriented Approach.* By Akbar K. Haghi, Ewa Klodzinska, Stefan Kubica, Gennady E. Zaikov (Eds.), Institute for Engineering of Polymer Materials and Dyes, Torun, p. 198 (2012).
14. *Selected nanopolymer research.* By A. K. Haghi, E. Klodzinska, S. Kubica, G. E. Zaikov (Eds.), Institute for Engineering of Polymer Materials and Dyes, Torun, p. 198 (2012).

4 A Study on Polymer Synthesis and Processing Using Supercritical Carbon Dioxide

A. V. Naumkin, A. P. Krasnov, E. E. Said-galiev,
A.Y. Nikolaev, I. O. Volkov, O. V. Afonicheva,
V. A. Mit', and A. R. Khokhlov

CONTENTS

4.1 INTRODUCTION

Medical grade Ultra-high-molecular-weight polyethylene (UHMWPE) has been widely used in medical applications due to its biocompatibility, high wear resistance, chemical inertness, very low friction, and high thermal stability. The main field of its medical application is total joint replacement (knee transplants). It is clear that, the surface properties of the prepared product are very important and, therefore, any surface treatment should be biocompatible as well. Development of work on the endoprosthesis of the joints caused, first and foremost successes in the study of ultrahigh molecular polyethylene. This polymer, since the 1962, remains the primary material used in hip and other joints. In recent years there has been an upsurge of research due to a modification of the structure of the polymer by physical methods. Improvement in tribological properties of the polymer providing the retention in biocompatibility is an important and unsolved problem up to date.

During the last 25 years, green chemistry has appeared and has vigorously developed [1]. The main objective in Supercritical carbon dioxide (SC-CO$_2$) treatment of this study was to improve the mechanical performance of UHMWPE products by means of formation of nanoporous structure and injection of CO_2. Surface modification with SC-CO_2 treatment is especially well suited for biomedical applications. The CO_2 is clean solvent, diluent, modifier for the synthesis and processing of a range of materials. Improvement of the properties is based on formation of porous structure in monolithic UHMWPE samples for the subsequent use as a "depot" for functional additivities. Modification of UHMWPE with SC-CO_2 treatment has been developed, and the original method for formation of porous structure in the polymer was suggested [2].

The XPS is a unique method for investigation of surface properties of materials, as it provides data on both the chemical state and the concentration of elements. However, during measurements of photoelectron spectra, at the surface of non-conducting samples, as a rule, positive charge is accumulated as a result of secondary electron emission induced by X-ray radiation [3, 4]. The charge shifts photoelectron peak to higher binding energy. In some cases the peaks are broadened. There are some ways to take into account such effects. Usually, as an internal reference the C 1s peak assigned to C-C/C-H bonds is used for charge compensation and it is positioned at 284.6–285.2 eV range. Sometimes a curve-fitting procedure is needed to distinguish C-C/C-H state in the C 1s spectrum. However, this procedure has a great deal of uncertainty in connection with a wide range of binding energies of different CH_xO_y groups [5]. To reduce the uncertainty one may use correlated fitting both C 1s and O 1s spectra. In this work, in accordance with [5], we use energy 285.0 for charge reference.

In a case of inhomogeneous and nonuniform samples, one part of sample can assumes different sample charge from other parts of the sample. The origin of this effect is regions with different secondary electron emission coefficient (g) and electrical conductivity (r) [6-10]. In our case we use differential charging as evidence of physical and chemical inhomogeneity and nonuniformity in original UHMWPE and treated with SC-CO_2. The change of surface elemental and chemical compositions has been studied by XPS in routine and differential charging modes. The advantages of differential charging to distinguish between chemical and physical features have been demonstrated. The approach based on comparison between C 1s and O 1s line shapes and their mutual positions. Both transport of secondary electrons, which depends on positive charge accumulated in the sample investigated, and bias polarity and voltage should be taken into account as well. At $U_b > 0$ a part of secondary electrons can move to the sample holder, while at $U_b < 0$ holes can. As a consequence charge redistribution in the bulk and surface can occur. It is evident, that charge distribution at the photoelectron information depth and related to it spectral line width depend on intensities of opposite flows from the X-ray gun foil and sample holder.

In this work we report a detailed XPS study of the influence of SC-CO_2 treatment on UHMWPE surface composition. Characterization of XPS data is based on differential charging, which reflects inhomogeneity of physical properties of the polymer. In this case, the inhomogeneity related to presence of oxygen, crystallinity, and porous structure. Oxygen may be in molecular state as H_2O, O_2, CO_2 and/or in bonded state as CH_xO_y groups.

4.2   EXPERIMENTAL

X-ray photoelectron spectra were recorded on a XSAM-800 Kratos spectrometer. A magnesium anode with characteristic radiation energy Mg Ka 1253.6 eV was used as the excitation source. The power deposited at the anode during recording the spectra did not exceed 90 W. Each spectral line was approximated by a Gaussian profile or the sum thereof and the background caused by secondary electrons and photoelectrons that lost energy was approximated by a straight line. The measurements were carried out at $\approx 10^{-7}$ Pa. The samples were secured to a titanium holder by two sided adhesive strip. The spectra were recorded at room temperature. The spectra were calibrated against C 1s line taking the binding energy corresponding to the C–(C,H) bonds to be 285.0 eV. Quantitative analysis was carried out using sensitivity factors of elements by a previously reported procedure [11].

4.3   DISCUSSION AND RESULTS

Figure 1 presents the survey spectrum of an original powder sample. The spectrum demonstrates photoelectron C 1s and O 1s peaks and related to them C KVV and O KVV Auger peaks. No either peak has been found. Elemental composition determined by XPS using atomic sensitivity factors (ASF) is close to $C_{95.6}O_{4.4}$. It should be noted that quantitative estimations based on ASF are valid for homogeneous samples only. In our case, as it will be shown below, the samples investigated are inhomogeneous ones because of molecular oxygen, which cannot be presented inside top surface layers in ultrahigh vacuum at room temperature; however, it can be at subsurface region. Nevertheless, we can use such quantitative data for comparison between different samples. As alternative to ASF, results of curve fitting photoelectron spectra are considered as well. The latter are more reliable because relative peak intensities in same spectrum are rather independent on photoelectron inelastic mean free path, which together with photoionization cross-section determine ASF. In other words, relative intensities of peaks in C 1s spectrum decomposition will be used for evaluation of oxygen content. The content of molecular oxygen, which is not exhibited in C 1s spectrum, will be determined by the same approach using decomposition of O 1s spectrum. Taking into account location of molecular oxygen one can expect that values of oxygen concentration will be less than actual ones.

Figure 2 shows high-resolution C 1s photoelectron spectrum, which exhibits a complex structure. In accordance with reference data [5] the C 1s spectrum was fitted with four peaks. The main one is assigned to polyethylene -CH_2-CH_2- backbone, while two smaller at 286.5 and 287.8 eV assigned to C-O/C-OH and C=O bonds, respectively. The fourth peak at 283.5 eV can be assigned to low-molecular weight species [12], which differ in g and r in comparison with macromolecules and charge to less potential by X-ray-induced electron emission. Their relative intensities are 89, 4, 5, and 2% respectively. Because overlapping ranges of chemical shifts in C 1s spectra for C-O-C and C-OH states [5], it is difficult to distinguish them. If we assume that only the C-OH state is at the surface, then the concentration of oxygen bonded to polymer carbon is ~7%, which significantly exceeds the value obtained with ASF. In opposite case of C-O-C state, the estimated value is ~4% that is close to value calculated with

ASF. On other hand, deconvolution of the corresponding O 1s spectrum (Figure 2) does not show presence of hydroxyl group and that only ~42% oxygen directly bonded to carbon.

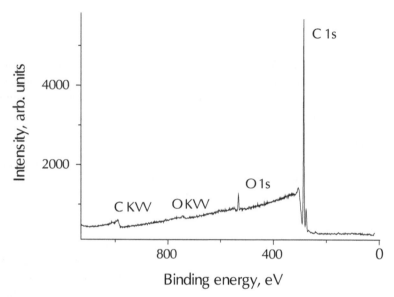

FIGURE 1   A survey spectrum of original UHMWPE powder sample.

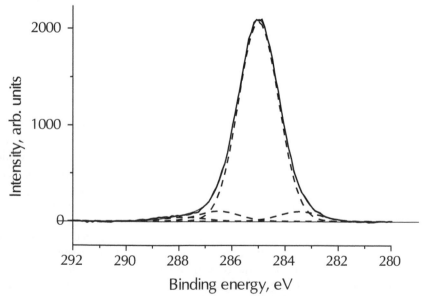

FIGURE 2   Deconvolution of the C 1s spectrum of original powder recorded at Ub = 0 V.

Figure 3 shows the deconvolution of O 1s peak of UHMWPE powder. Two peaks in fitting the O 1s spectrum at 532.6 and 532.3 eV are related to the C 1s spectrum and assigned to C-O-C and C=O bonds, respectively. It was seen that two additional peaks should be used to adequate fitting, namely at 533.6 and 531.3 eV. The first component is assigned to H_2O, while the second one cannot be assigned to any bond between carbon and oxygen atoms [5]. Binding energy of 531.3 eV is typical for metal oxides [13], however, neither photoelectron nor Auger peaks of metals are observed in the survey spectrum. Therefore, they should be assigned to molecular oxygen species such as H_2O or O_2. Thus, in subsurface region oxygen is in solid and gaseous phases. For further consideration, the spectra recorded with bias voltage applied to the sample holder will be analyzed.

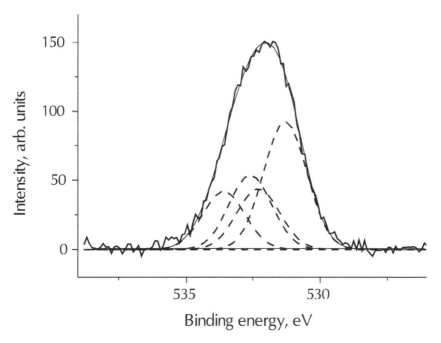

FIGURE 3   Deconvolution of the O 1s spectrum of original UHMWPE powder recorded at $Ub = 0$ V.

Figures 4 and 5 compares the C 1s and O 1s spectra of original UHMWPE powder, recorded at different bias voltage applied to the sample holder.

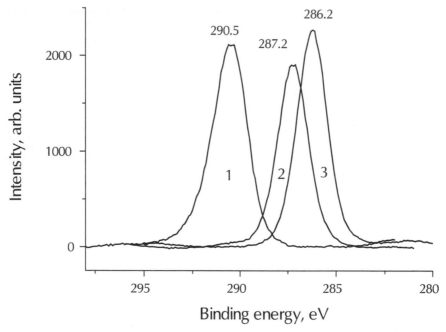

FIGURE 4 The C 1s spectra of original powder recorded at Ub = 7 (1), 0 (2) and –7 V (3), without compensation of surface charging.

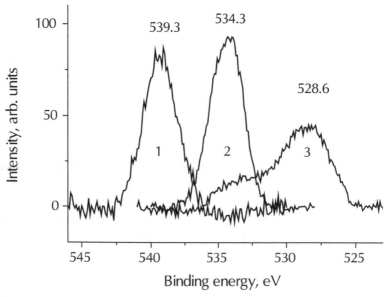

FIGURE 5 O 1s spectra of original powder recorded at Ub = 7 (1), 0 (2) and –7 V (3), without compensation of surface charging.

The O 1s region recorded at $U_b = -7$ V showed a clearly resolved two-peak structure, the left one being C-O bonds at ~534 eV, and the second unidentified one at 528.6 eV. It is clearly seen that both the C 1s and O 1s spectra recorded at $U_b = -7, +7$ V are shifted by different values (Δ_1, Δ_2) relative to the spectra recorded at $U_b = 0$ V, because conductivity in subsurface region depends on flooding with secondary electrons from X-ray gun foil, which is used to cut off bremstrahlung radiation. At $U_b = 7$ V conductivity the flow of secondary electrons is more than that at $U_b = 0$ V, while the second one is more than that at $U_b = -7$ V. At $U_b = -7$ V the most part of secondary electrons is cut off.

It should be noted that the C 1s and O 1s spectra are shifted by different values (Δ_1 and Δ_2) when bias voltage is applied. It means that mainly carbon and oxygen atoms are located in regions differed in physical properties. Δ_1 and Δ_2 values indicate that regions containing oxygen have better conductivity than those containing carbon. Δ_{1O} and Δ_{2O} values differ by 0.5 eV, while Δ_{1C} and Δ_{2C} differ by 2.3 eV, that is shift of the O 1s peaks slightly depends on bias polarity. This means that corresponding O 1s peaks are related to gaseous molecules. On other hand, in the O 1s spectrum recorded at $U_b = -7$ V, the second broad peak at ~533.0 eV is characterized with Δ_O ~ 1.2 eV, which is close to Δ_{1C} ~ 1.0 eV, and this peak can be attributed to oxygen bonded to carbon.

For better understanding the results obtained we have used comparison between the C 1s and O 1s line shapes. For this purpose the O 1s spectra were overlaid with the corresponding C 1s spectra as shown in Figure 6. All the C 1s spectra were shifted by the same value to achieve maximum overlapping of them. This procedure allows directly discriminate relative concentrations of oxygen atoms bonded/non-bonded to backbone carbon.

The Figure 6 supports results obtained with deconvolution the O 1s spectum presented in Figure 3. The same conclusion can be obtained after compensation of surface charging (Figure 7). Both Figures 6 and 7 shows that a minor part of the O 1s peak area (~13%) is related to oxygen atoms bonded to backbone carbon.

It is often assumed that the criterion for compensation of surface charging, when electron flood gun is used, is a minimum width of photoelectron peak, which is a consequence of leveling the surface potential. However, this criterion is no applicable if sample is grounded and contains insulating and conducting regions. In the case studied, increase of secondary electron flow from the X-ray gun foil at $U_b = 7$ V does not lead to a decrease of peak width in comparison with that recorded at $U_b = 0$ V. This indicates that instead of leveling the surface potential is nonuniform charge distribution caused by the "opening" of channels to drain the charge as a result of the induced conductivity. Apparently, these channels appear at the surface regions containing oxygen, because minimum width of O 1s peak is observed at $U_b = 7$ V. It may also mean that the charge carriers are generated ions of molecular oxygen or water.

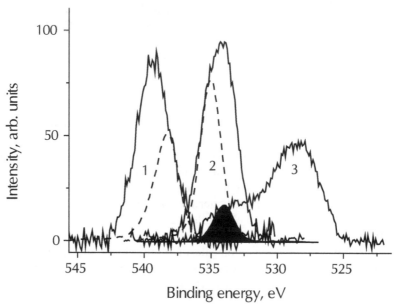

FIGURE 6   C 1s and O 1s spectra of original powder recorded at U_b = 7 (1), 0 (2) and –7 V (3) after compensation of surface charging. The shaded area shows relative intensity of oxygen atoms bonded to backbone carbon.

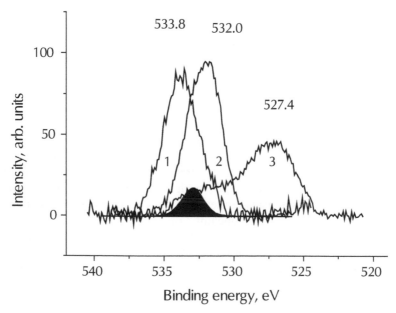

FIGURE 7   O 1s spectra of original powder recorded at Ub = 7 (1), 0 (2) and –7 V (3) after compensation of surface charging. The shaded area shows relative intensity of oxygen atoms bonded to backbone carbon.

At $U_b = -7$ V, in contrast to case of $U_b = 7$ V, secondary electron flow from X-ray gun foil decreases significantly causing decrease in induced conductivity, which in turn levels surface potential in regions with different conductivity by "closing" channels to drain the charge. This contributes to narrowing the C 1s peak. Using a similar approach to the O 1s spectra one can conclude that regions containing oxygen bonded to backbone and gaseous oxygen have different surface potential because drain channels are closed. This contributes to broadening the O 1s peak. It should be noted, that in the C 1s spectrum recorded at $U_b = -7$ V, the state observed at $U_b = 0$ V at binding energy of 287.8 eV and assigned above to C=O bond is disappeared. Instead of it a peak at 280.2 eV has been found.

In the C 1s spectrum recorded at $U_b = 7$ V, there are all the states observed at $U_b = 0$ V and one additional state at 290.4 eV. Therefore the peaks discussed should be attributed to differential charging of some regions, change of which potential is close to bias voltage. This consideration shows that using a routine identification of photoelectron spectra it is easy to do false conclusions.

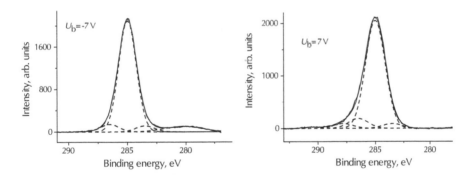

FIGURE 8   Deconvolution of the O 1s spectra of original powder recorded at Ub = –7 V and 7 V.

Figure 9 shows another approach to interpretation of the C 1s spectra, based on comparison charge reference and normalization by area. The C 1s spectra, normalized by area, show that only overlapping part can be used for reliable identification of chemical states. Other non-overlapping part is resulted from differential charging and reflects carbon species with other physical properties, namely g and r parameters. The species can be assigned to surface contaminations.

Survey spectra of UHMWPE recorded after SC-CO$_2$ treatment at 65, 72.5, 85, 92.5, and 100°C with and without preliminary CO$_2$ purge, from qualitative point of view, are similar but evidence different oxygen content of 4.4, 4.8, 2.0, 6.4, 3.9, and 4.4%, respectively. According to results obtained for initial UHMWPE powder we will applied the most simple and reliable approach to identify peculiarities, observed in the C 1s and O 1s spectra of UHMWPE samples after SC-CO$_2$ treatment at 65, 72.5, 85, 92.5, and 100°C with biasing sample holder. Therefore, below the C 1s and O 1s spectra recorded at $U_b = -7$ V will be considered as the most informative ones. As depicted in Figure 10 for the SC-CO$_2$ treated polymer, the recorded line shape and relative intensity of oxygen bonded to backbone carbon are strongly dependent on temperature. At 72.5°C appears a double peak structure, and further the peak at low-energy side shifts to lower binding energy reaching the maximum value o5 7.5 eV at 85°C and then moves backward. In other words the temperature of 85°C is a singular one. Evolution of the O 1s spectra is interpreted as embedding of CO_2 in nanopores formed with SC-CO_2 treatment. However, reliable quantitative estimation of CO_2 content is in progess. As a matter of fact we have found segregation of some Si-containing species since 72.5°C. Based on evolution of Si content and intensity of C 1s peak at ~280 eV we assign it to siloxane group. It was found that a maximum of Si content is 3.1% at 92.5%. It should be noted that after treatment at 85°C Si peaks have not been observed in photoelectron spectra. One can speculate that CO_2 remove siloxane out of polymer.

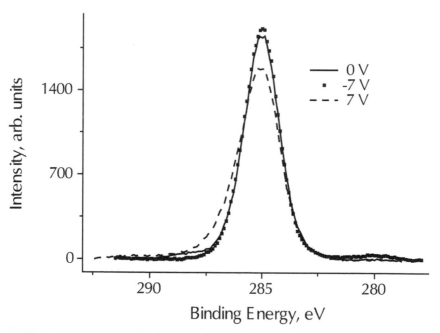

FIGURE 9   C 1s spectra of original powder recorded at Ub = 7, 0 and –7 V after compensation of surface charging. The spectra are normalized by area.

It was of interest to examine effect of surface contamination on photoelectron spectra. For this purpose a polymer sample was pre-purged with SC-CO_2 at room temperature and then was treated at 100°C. Under these conditions, double-peak structure in the O 1s spectrum recorded at $U_b = -7$ V observed after treatment at other temperatures disappeared and H_2O content decreased.

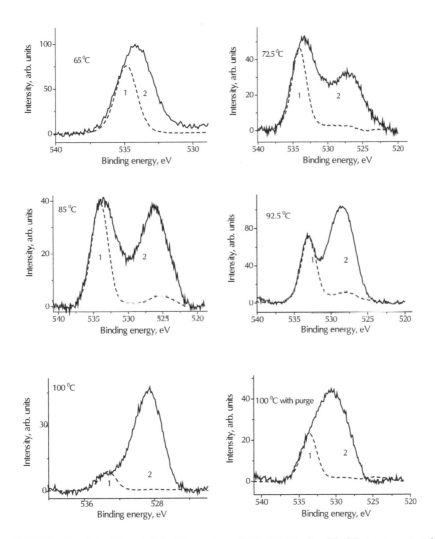

FIGURE 10   C 1s (1) and O 1s (2) spectra of UHMWPE after SC-CO$_2$ treatment at 65, 72.5, 85, 92.5, 100°C (with/without purge) recorded at $U_b = -7$ V.

Results of friction tests performed on a friction machine of type-I47 indicate applicability of XPS to solving tribological problems of UHMWPE. Figure 11 shows temperature dependence of frictional heating of UHMWPE plate on the treatment temperature with SC-CO$_2$. One can see the gradual decrease of frictional heating with increasing temperature SC-CO$_2$ till ~85°C, and as it follows from the XPS data, the sharp rise since 100°C.

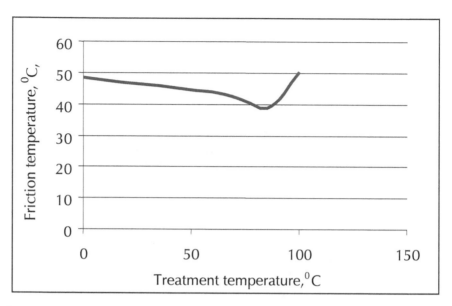

FIGURE 11   Dependence of the friction temperature on temperature of SC-CO$_2$ treatment of UHMWPE.

This can be attributed to the distinctive surface properties, detected by XPS, which are mainly expressed in the presence of gaseous CO$_2$, which has remained in the nanopores. Apparently, the temperature at 65°C is not sufficient for the formation of a large number of nanopores, while the temperature of 100°C leads to a change in the morphology of the polymer, and "openings" nanopores. Also it was found that treatment at 85°C reduced surface oxidation with friction (Table 1).

TABLE 1   Surface composition of UHMWPE samples determined by XPS.

Sample	65 °C	85 °C
original powder	4.4%	
treated powder	2%	2.6%
plate after friction	6%	5%

Comparison of the deconvoluted O 1s spectra of polymers treated at 85 and 72.5°C shows that amount of CO$_2$ is higher in the first sample.

4.4   CONCLUSION

These results demonstrate the advantages of using controlled differential charging in XPS measurements to study modification of friction surface of polymer tribological materials. The appearance of gaseous CO$_2$ in the subsurface layer of UHMWPE after exposure to SC-CO$_2$ can be due to formation of nanopores and encapsulation of

CO_2 in them. The approach based on differential charging and comparison between C 1s and O 1s line shapes and their mutual positions is applicable to a wide class of non-conducting polymers due to its simplicity and reliability to distinguish physical inhomogeneities at polymer surface.

KEYWORDS

- **Atomic sensitivity factors**
- **Photoelectron spectra**
- **Quantitative analysis**
- **Spectrum**
- **Supercritical carbon dioxide**
- **Surface composition**
- **Ultra-high-molecular-weight polyethylene**

REFERENCES

1. Cooper, A. I. Polymer synthesis and processing using supercritical carbon dioxide. *J. Mater. Chem.*, **10**, 207 (2000).
2. Naumkin, A. V., Krasnov, A. P. Said-Galiev, E. E. Volkov, I. O., Nikolaev, A. Yu., Afonicheva, O. V., and Khokhlov, A. R. Carbon dioxide in the surface layers of ultrahigh molecular weight polyethylene. *Doklady Physical Chemistry*, **419**, 68 (2008).
3. Cazaux, J. Mechanisms of charging in electron spectroscopy. *J. Electron Spectrosc. Relat. Phenom.* **105**, 155 (1999).
4. Veereecke, G. and Rouxhet, P. G. Surface charging of insulating samples in x-ray photoelectron spectroscopy. *Surf. Int. Anal.*, **26**, 490 (1998).
5. Beamson, G. and Briggs, D. *High Resolution XPS of Organic Polymers: The Scienta ESCA 300 Database.*, Wiley, Chichester (1992).
6. Tielsh, B. J. and Fulghum, J. E.: Differential charging in XPS. Part III. A comparison of charging in thin polymer overlayers on conducting and non-conducting substrates. *Surf. Int. Anal.*, **25**, 904 (1997).
7. Havercroft, N. J. and Sherwood, P. M. A. Use of differential surface charging to separate chemical differences in x-ray photoelectron spectroscopy. *Surf. Int. Anal.*, **29**, 232 (2000).
8. Ertas, G. and Suzer, S. XPS analysis with external bias: a simple method for probing differential charging. *Surf. Int. Anal.*, **36**, 619 (2004).
9. 9. Suzdalev, I. P., Maksimov, Yu. V., Vasil'kov, A. Yu., Naumkin, A. V. Podshibikhin, V. L. and Volkov, I. O. Electronic and Magnetic Properties of Au-Fe Cluster Nanocomposites Prepared by a Sequential Solvated Metal Atom Dispersion Process. *Nanotechnologies in Russia*, **3**, 72 (2008).
10. Vasil'kov, A. Yu., Naumkin, A. V., Volkov, I. O., Podshibikhin, V. L., Lisichkin, G. V. and Khokhlov, A. R. Surf. *Interface Anal.* **42**, 559 (2010).
11. Naumkin, A. V., Volkov, I. O., Tur, D. R., and Pertsin, A. I. X-ray Photoelectron Spectroscopic Analysis of the Polyphosphazene Surface. *Polymer Science*, **B44**, 139 (2002).
12. Grüneis, A., Kummer, K., and Vylykh, D. V. Dynamics of graphene growth on a metal surface: A time-dependent photoemission study. *New J. Physics*, **11**, 073050 (2009).
13. Wagner, C. D., Naumkin, A. V., Kraut-Vass, A., Allison, J. W., Powell, C. J., and Rumble, J. R. Jr. NIST Standard Reference Database 20, v. 3.4, Web Vers., (2000) http://srdata.nist.gov@xps.

5 Updates to Production and Application of Chlorine Polyolefine

Yu. O. Andriasyan, I. A. Mikhaylov,
A. A. Popov, A. L. Belousova, A. E. Kornev,
and Yu. G. Moskalev

CONTENTS

5.1 INTRODUCTION

Based on historical data halide modification (HM) of high molecular compound was carried out in 1859, natural rubber (NR) was exposed to modification and, in addition to that, NR was dissolved in perchloromethane, through which chlorine gas was run through. Modified NR is powder product with content of fixed chlorine not over 6268% m., which did not have properties of elastomer [1, 2]. The HM of NR may be referred to one of the first attempt of commitment of new properties to polymer with help of carrying out of chemical modification.

Nowadays HM of polymers together with obtaining of halogen-containing polymers with help of synthesis is one of intensively developing direction in the field of obtaining chlorine-containing polymers. In result of carrying out of HM of polymers, which have technologically smoothly, large capacity industrial production, elastomer materials, and composites are managed to obtain with wide complex of new specific properties: high adhesion, fire oil, gasoline, heat resistance, ozone resistance, incombustibility, resistance to influence of corrosive environments, microorganisms, high strength, gas permeability and so on.

Nowadays by world polymer industry was developed manufacture of those widespread polymers of HM, which has properties of elastomers such as: Chlorosulfonated polyethylene (CSP), chlorinated polyethylene (CP), chlorinated and bromated butyl rubber (CBR, BBR) and chlorinated ethylene-propylene (CEP), and ethylene-propylene-diene cauotchoucs (EPDC) in small amount.

In this chapter we consider questions, concerning with obtaining, and processing of halide modified chlorine-containing cauotchoucs as CBR and CEPC, which are prospective in terms of application in rubber industry. Perceptivity of their production and application consists in specific properties of these cauotchoucs (high gas permeability of CBR and high heat, ozone resistance of Chemical Engineering Process Design Centre (CEPDC)). These properties are caused by structure of both initial (BR and EPDC) and chlorine-containing cauotchoucs (CBR and CEPDC).

Originally, before carrying out of HM of cauotchoucs BR and EPDC, attempts of rubber application based on these cauotchoucs were undertaken for purpose of items creation from elastomer materials, differing in high gas permeability and high heat ozone resistance. In the process of properties study of rubber mixtures and rubbers from these cauotchoucs was found, that rubber mixtures had unsatisfactory characteristics by manufacturability of obtaining and processing. For the purpose of improvement of technological characteristics of rubber mixtures, attempts of combination of cauotchoucs BR and EPDC with diene cauotchoucs (NRs and synthetic isoprene rubber (SIR-3) etc) was undertaken. However, this combination did not result in positive decision of given problem. If production and processing of rubber mixtures based on combined system of cauotchoucs with technological point of view did not provoke difficulties, that creation of rubber items, which are able to use, is impossible. The reason is that if we combinate cauotchoucs, which differ in its unsaturation in case of application of, sulfur vulcanization that resulted in absence of covulcanization between phases of combined cauotchoucs [3]. Thus, there was no unified, spatial, vulcanized network in rubbers based on combined system. In process of vulcanization took place redistribution of catalyst with help of diffusion and vulcanizing agent from the phase of cauotchouc with less unsaturation (BR and EPDC) to the cauotchouc's phase with high unsaturation (NR and SIR-3). Obtained rubbers are not satisfied with its strength and dynamic characteristics.

Many decisions of this problem were suggested but the most effective was decision to add trace amount of halogen in macromolecular structure of cauotchouc with low unsaturation [4, 5]. It gave additional functionality to cauotchouc and therefore, higher vulcanization rate. The optimum halogen content was when deterioration of initial cauotchoucs specific properties was not observed, and additionally capability of halide-containing cauotchouc to be covulcanized with high unsaturated elastomers was gathered.

Historically, the most popular in tire industry was CBR. The CEPDC cauotchouc had limited application, because required level of rubbers ozone resistance in rubber technology, traditionally, created by adding of chemical age resistors and antioxidants. Rubber ozone resistance and service time of the rubber item had the same duration. It is necessary to note, that this protect is inefficient for items with long period of service, because of exudation of age resistors and antioxidants from rubber. It is important to note, that the fraction of this items in common amount of output rubber products is very insignificant.

There were no problems on the first industrial production stage of HM of chlorine-containing cauotchoucs, the requirement in this cauotchoucs was growing, that was promoting to open new factories for production these cauotchoucs. However, it is necessary to note, that since realization of HM of NR in 1859 almost nothing was changed in technology of obtaining of chlorine-containing cauotchoucs. This technology was preserved with some small changes until the present time. The meaning of given technology [6] or as it called by specialists "dissolved technology" is that, on the first stage, the polymer, which we want to modify, is dissolved in organic solvent. The concentration of solution should not exceed 10% from technological consideration. Then gaseous halogen, it is chlorine or bromine, is leaked through obtained solution, then when planned content of halogen in polymer is reached, the process is suspended. Then are following the stages: detrainment of obtained chlorine-containing polymer, its washing and neutralization, then the stage of drying, packing and storage. As secondary process we can consider recuperation of solvent. All developments of this technology consisted in replacement of gaseous halogen with halogen-containing organic compounds, that did not promote simplification of both technology and ecology of production process. Dissolved technology of obtaining of chlorine-containing cauotchoucs is multistage process, which in terms of current, strict ecology requirements does not stand up to scrunity.

Taking into account the above disadvantages of dissolved technology of obtaining of chlorine-containing cauotchoucs, alternative technology of obtaining of chlorine-containing cauotchoucs was developed and was offered in the end of ninetieth of past century by scientists' community and specialists of Moscow Academy of Fine Chemical Technology, Institute of biochemical physics, research, and manufacturing association of firms "Polikrov" and "The Moscow tire factory" [7]. The differential characteristic of new technology is technological simplicity of carrying out obtaining process of chlorine-containing cauotchouc and its ecological safety.

The developed technology is based on solid phase (mechanochemical) HM of initial cauotchoucs by chlorine-containing organic compounds, which are environmentally safe in process of carrying out of HM. The developed technology has patent protection and opportunity to obtain both CBR and CEPDC, and others (saturate and unsaturated) cauotchoucs. Within the framework of newly developed technology is assimilated experimental industrial output of cauotchoucs CBR-2,5 and CEPDC-2,0 (the number shows content of fixed chlorine in cauotchouc).

5.2 DISCUSSION AND RESULTS

Research industrial testing of cauotchouc CBR-2,5 in rubber formula of radial tires inner lining, tubeless construction, was carried out on the Moscow tire factory. The point of carrying out of investigations was to substitute serially used rubbers of inner lining of chlorine-containing cauotchouc HT-1066 (produced by USA) for cauotchouc CBR-2,5. Conducted investigations showed, that production and processing of rubber mixtures with new cauotchouc CBR-2,5 on technological equipment created no problems. Plasto-elastic, physical-mechanical, and some specific properties of serial and experiment rubber mixtures and their vulcanizates, containing cauotchouc CBR-2,5 were studied. The results of investigations are showed in Table 1.

TABLE 1   Properties of rubber mixtures and rubbers for radial tires' inner lining with application of serial chlorine-butyl rubber CBR HT-1066 and CBR-2,5.

Index	HT-1066	CBR-2,5
Plasticity	0.37	0.40
Cohesion strength, MPa	3.49	3.45
Mooney viscosity(100°C)	58.5	66.0
Plastometer "Faerston" tests		
Flow time of rubber mixture, s.	25,8	16.2
Shrinkage,%	62.0	58.5
Monsanto rheometer tests		
Rotational moment, N*m		
Min	9.0	9.8
Max	16.0	24.5
Initial time of vulcanization, min	4.4	9.3
Vulcanization rate, %/min	7.9	9.4
Optimum vulcanization time, min	17.0	20.0
Physical-mechanical indexes		
Conventional tensile strength 300%, MPa	4.2	6.9
Conventional tensile strength, MPa	10.5	10.0
Conventional breaking elongation, %	650	550
Tear resistance, kN/m	31	39
Gas permeability (to hydrogen), $l/(m^2 * d)$	0.49	0.52

We can see from the Table 1 that experimental and serial rubbers almost did not differ in plasticity, mooney viscosity, and cohesion strength.

Test of rubber mixtures on plastometer "Faerston" found higher fluidity of experimental rubber with cauotchouc CDR-2,5.

Study of vulcanized characteristics of the rubber mixtures on Monsanto rheometer showed, that experimental mixtures with cauotchouc CDR-2,5 excel serial mixtures almost in two times in the initial vulcanization time and have higher vulcanization rate in basic period, that is very important with technological point of view.

Study of physical-mechanical characteristics of the rubbers showed, that experimental rubber much excel serial in tensile strength 300%, but there are no differences between experimental and serial rubbers in strength, conventional breaking elongation, and tear strength.

The values of gas permeability (to hydrogen) of experimental and serial rubbers are the similar.

Thus, in the Table 1 we showed, that new chlorine-containing butyl cauotchouc CBR-2,5 satisfies the requirements by their characteristics, demanded on halogen-containing butyl cauotchoucs, used in rubber production of inner lining.

The next stage of our investigations was to study opportunities of application of new chlorine-containing cauotchouc CEPDC-2 in the formulas of rubbers for sidewall of radial tires and rubbers for production of diaphragm press.

As we know, sidewalls' rubber is exposed deformations in process of service that is the reason of intensive heat emission. Increased temperature promotes premature heat and ozone ageing of rubbers of tire sidewalls. Traditionally, chemically synthesized antioxidants and age resistors are mixed in rubber formula for protection of sidewall rubber from heat and ozone ageing [8]. The "bleeding" of protectors takes place in process of service, because they do not bind chemically with elastomer matrix; all this reasons promote premature ageing and destruction of sidewalls. Considering that part of tires can be reconstructed after service period, it will be very practically advantageous to increase heat and ozone resistance of sidewalls by adding of protection component, which can build into elastomer matrix with help of chemical links. The function of that component can make new chlorine-containing cauotchouc CEPDC-2, because we know, that it has capability to covulcanize with high unsaturated cauotchoucs, composing on rubber formula for sidewalls. It is well-known, that adding 2030 mass part of cauotchouc CEPDC-2 is enough for increase of ozone resistance of rubber from diene cauotchoucs [9].

In this case we studied the opportunity of application of cauotchouc CEPDC-2 in rubber formula for sidewalls of radial tires, elastomeric part of which has diene cauotchoucs SIR-3 and CDR in ratio (50:50). The ratio of cauotchoucs SIR-3:CDR:CEPDC-2 was 50:20:30 and 50:30:20 in experimental rubber. Chemical antioxidants were not added in experimental rubber mixtures.

We established that production and processing of rubber mixtures with cauotchouc CEPDC did not have difficulties on technological equipments. We studied plasto-elastic, physical-mechanical and some specific properties of serial and experiment rubbers. Experimental data are showed in Table 2.

TABLE 2 Properties of studied serial and experimental rubber mixtures and rubbers, based on cauotchoucs SIR-3, CDR, and CEPDC-2.

Index	Serial rubber*	Experimental rubber	
		1**	2***
Mooney viscosity(120°C)	43	45	47
Plasticity	0.44	0.44	0.36
Conventional modulus at 300%, MPa	3.7	7.2	6.0
Conventional tensile strength, MPa	15.7	20.5	18.4

TABLE 2 *(Continued)*

Index	Serial rubber*	Experimental rubber	
		1**	2***
Conventional breaking elongation, %	770	600	610
Conventional permanent tension elongation, %	14	15	13
Coefficient of heat ageing (100°C, 72h)			
At strength	0.56	0.85	0.82
At conventional elongation	0.63	0.92	0.91
Coefficient of ozone resistance of dynamic tests ($\varepsilon = 20\%$)	0.52	0.95	0.92
TM-2 hardness	56	60	62
Rebound elasticity, %			
under 20°C	41	42	44
under 100°C	47	50	52
Crazing strength, th. cycle	>252	>252	>252
Dynamic repeated tension durability, th. cycle	>50	>50	>50

*Based on cauotchoucs SIR-3 - CDR (50:50),
**Based on cauotchoucs SIR-3 – CDR – CEPDC-2 (50:30:20),
***Based on cauotchoucs SIR-3 – CDR – CEPDC-2 (50:20:30).

We can see from the Table 2 that plasto-elastic characteristics of serial and experimental rubbers have close values; serial rubbers have conventional modulus at 300% twice higher as experimental ones and have higher conventional tensile strength and hardness. The value of rebound elasticity, crazing strength, and dynamic repeated tension durability of experimental and serial rubbers are very similar. It should be noted, that serial rubbers have heat resistance and ozone resistance doubles that serial rubbers, containing antioxidants.

Thus, the investigations showed that new chlorine-containing cauotchouc CEP-DC-2 in rubber formulas for tires sidewalls can be used as polymer antioxidant.

The practice states that the main reasons of breakdown of diaphragm press are the low capacity to elastic recovery of rubbers based on butyl cauotchoucs, leading to "treading out" of diaphragm, and high extent of "tar value" of diaphragm work surface. To eliminate these disadvantages we studied opportunity of substitution of cauotchouc SEPC-60 in formulas of serial rubbers (resin curing) for diaphragm for new chlorine-containing ethylene-propylene-diene cauotchoucs CEPDC-2.

Cauotchoucs BR-1675 and SEPC-60 in ratio (85:15) compose formula of serial rubbers for diaphragm production, in experimental rubber SEPC-60 was substituted for similar amount of CEPDC-2. It is well known, that chlorine-containing compounds have the ability to activate resin curing of butyl cauotchouc, which is the main elastomer component of membraneous rubbers [1].

The investigations showed that there were no difficulties in production and processing of rubber mixtures with cauotchouc CEPDC-2 on technological equipments.

We studied plasto-elastic, physical-mechanical, and some specific characteristics of serial and experiment rubber mixtures and rubbers. Experimental data are showed in Table 3.

TABLE 3 Properties of serial and experimental rubber mixtures and rubbers for production of shaper-vulcanization.

Index	Serial rubber	Experimental rubber
Plasticity	0.41	0.42
Mooney viscosity(140°C)	37	36
Conventional modulus at 300%, MPa	5.0	6.0
Conventional tensile strength, MPa	10.2	12.6
Conventional breaking elongation, %	620	600
Conventional permanent tension elongation, %	34	20
Tear resistance, kN/m	60	63
Coefficient of strength heat ageing (180°C, 24hr)	0.6	0.6
Coefficient of strength temperature resistance under 100°C	0.7	0.62
TM-2 hardness	74	78
Dynamic repeated tension durability (ε_{dyn} = 50%; εsta t = 37.5%), th. cycle	42	>50
Creep (160°C, 24hr, 0.3M Pa),mm	119	53
Rebound elasticity*, %	13/18	18/32
Rebound elasticity*, after ageing, %	16/28	18/30
Tar value, %	1.2	0.6

*In numerator under 20°C, in denominator under 100°C

We can see from the Table 3, that plasticity and Mooney viscosity of experimental and serial rubber mixtures have close values. The values of conventional modulus at 300%, conventional tensile strength, tear resistance, hardness, rebound elasticity, and dynamic durability of experimental rubbers are higher than serial ones.

It should be noted, that experimental rubber has lower by half conventional permanent tension elongation in comparison with serial rubber, although the values of conventional tensile elongation are the similar; "tar value" and creep under 160°C are lower (in two times and more, than in two times respectively).

5.3 CONCLUSION

Thus, we can make conclusion, based on obtained results that application of new chlorine-containing cauotchouc CEPDC-2 in rubber formula for production of press diaphragm will permit to increase the diaphragm service time.

It should be noted, that developed new technology of obtaining of chlorine-containing cauotchoucs permits to manufacture competitive chlorine-containing polyolefine cauotchoucs CBR-2,5 and CEPDC-2. As showed the investigations of cauotchouc CBR-2,5, well recommended yourself in conditions of rubber production and cauotchouc CEPDC-2, which did not have analogues on synthetic cauotchoucs market, can be used as polymeric antioxidant in rubbers based on diene cauotchoucs.

KEYWORDS

- **Cauotchoucs**
- **High adhesion**
- **Incombustibility**
- **Mooney viscosity**
- **Natural rubber**

REFERENCES

1. Doncov, A. A., Lozovik, G., and Novizkaya, S. P. *Chlorined polymers. M. "Chemistry"*, p. 232 (1979).
2. Ronkin, G. M. Current state of production and application of chlorine polyolefine. M.: *NIITEChem*, p. 81 (1979).
3. Chirkova, N. V., Zaxarov, N. D., and Orexov, S. V. M. Rubber mixtures based on combination of cauotchoucs. p. 62 (1974).
4. Morrissey, R. T. Halogenation of Ethulene Propylene Diene Rubbers. *Rubber Chem. And Technol*, **44**(4), 10251042 (1971).
5. Ronkin, G. M. Investigation of chloring ethylene-propylene-diene copolymers process and properties of obtained modifications. *Manufacturer of SC.*, (6), 811 (1981).
6. Ronkin, G. M. Chlorosulfonated polyethylene. *M. TsNIITEneftekhim*, 101 (1977).

7. Andriasyan, Y. O. Elastomeric materials based on cauotchoucs, exposed mechanochemical halide modification. 05. 17. 06, *M.* 362 (2004).
8. Ragulin V. V. Production if pneumatic tires. *M. "Chemistry"*, 504 (1965).
9. Andriasyan, Y. O. Investigation of properties of rubber mixtures and vulcanizates based on combined system of unsaturated cauotchoucs with halogenated ethylene-propylene cauotchoucs. 05. 17. 12. *M.* 212 (1981).

6 A Study on Rheological Properties of Dilute and Moderately Concentrated Solutions of Chitosan

S. A. Uspenskiy, G. A. Vikhoreva, A. N. Sonina,
N. R. Kildeeva, L. S. Galbraikh, and
A. N. Kosygin

CONTENTS

6.1   INTRODUCTION

The preparation of chitosan films, pellets, fibers and other products is made by disolution chitosan. The dilute aqueous solutions of acetic acid commonly are using as the cheapest and most accessible reagent that provides the formation of water soluble polymer derivative chitosan acetate. However, the high boiling point of the solvent (>100°C) and relatively low concentration of the polymer in forming solutions (£6%) causes the low speed of the thermal spinning process.

There are various techniques to increase the speed of the solvent evaporation, for example the use of solvent's mixtures with different boiling points, as well as solvent mixtures with precipitator [1]. Despite the fact that it usually provides a "tough" conditions spinning and forming non-equilibrium and defect structure of the polymer products, in some cases the use of this method is justified. Thus, much attention is paid to the formation of porous structure in obtaining of separation membranes and fibrous sorbents, including bicomponent structure "core-shell".

They are few dates about the influence of alcohol on the rheological properties of aqueous solutions of polymers. It is known that the introduction of alcohol in aqueous solutions of polyvinyl alcohol and methylcellulose improves their dissolution and suppress gelation [2]. The introduction of ethanol in the spinning solutions of cellulose derivatives is widely used for reducing their viscosity and improves the homogeneity [3]. The presence of alcohol in acetic acid chitosan solutions influences on degree protonation of amino groups of polymer [4] that can be important for a number of practical applications, in particular for sorption or cross-linked chitosan. In works [5, 6] introduction methanol in to chitosan solutions was used for creation sfter conditions of polymer acylation and decreasing the viscosity of solutions without any investigation this problem.

The aim of the work is to investigate the influence of the ethanol introduction on structural and rheological properties chitosan's solutions in acetic acid, the rate of curing of solutions and the quality of chitosan membrane affixed on viscose filament to give it a sorption capacity. This chapter presents the results of a comparative study of properties of dilute and concentrated chitosan solutions in acetic acid in the absence and presence of ethanol. The choice of ethanol is due to a good compatibility with the main solvent, a lower boiling point than water, as well as its relative small toxicity. The used chitosan had MM $\sim 1.7 \times \times 10^5$, degree of deacetylation 0.92, degree of crystallinity $\sim 45\%$, 100% solubility in 2% acetic acid and $\leq 0.2\%$ of ash content.

6.2 EXPERIMENTAL

The solutions for rheological studies were prepared by conducting the preliminary swelling of chitosan in water, followed by dissolution in acetic acid and diluted alcohol, water-alcohol mixture and water to achieve a given polymer concentration (1–8 wt %) and the content of alcohol (10–40 wt %). Ethanol, water or water-ethanol mixtures are injected into the system in small portions with vigorous stirring to prevent precipitation of the polymer. Non-Newtonian chitosan's solutions are not stable and the viscosity of solutions gradually decreases (most strongly in storage during the first day [7-9]). Therefore, in comparative experiments assessed values of the solution viscosity, measured at a fixed value of shear stress (σ) or velocity gradient (j) 3 c^{-1}, after conditioning for 24–30 hr. The viscosity of chitosan's solutions essentially depends on acid concentration [10, 11], so to investigate their rheological properties there were used the same molar ratio of acid: chitosan, equal to 2.2, and the same degree of protonation of amine groups. Rheological curves of solution were measured with a Rheotest-2 rotary viscometer. The activation energy Ea was calculated for polymer solutions in the temperature range 20–60°C from viscosity values measured at t = 10 Pa.

Viscosity of solvents was measured in capillary viscosimeter with a known constant for the calculation of dynamic viscosity.

Characteristic viscosity of chitosan solutions was determined at 25°C in the viscometer Ubbelode with capillary diameter 0.56 mm to identify the effect of ethanol on the thermodynamic quality of solvent, and, consequently, the conformation and the degree of macromolecules association, the ionic strength was created by minimal amount of sodium acetate (0.04 mol/l) that subsided polyelectrolyte swelling. The concentration of acetic acid and the initial concentration of chitosan solution are 0.4%. The optical density (D) of 1% solutions was determined on a spectrophotometer UNICO 1200/1201 in the wavelength range (l) at 350–900 nm with intervals of 40 nm. The calculation of solution's turbidity (t), number (N) and radius (r) of scattering particles carried in the range 420–600 nm, where the difference of optical density of chitosan solutions with and without alcohol is the most significant according to formulas [12]:

$$t = 2.3D_{cp}/l$$

$$N = 1.26 \times 10^{17} t/l_{cp}^2 \, \alpha^2 k$$

$$r = l_{cp} \, a/2 \times \times 3.14 n_0$$

where l is длина кюветы cm, a and k = characteristic constants of light scattering [12], n_0 - refractive index of the solvent.

6.3 DISCUSSION AND RESULTS

Since ethanol is a precipitator for chitosan, its introduction into solutions of chitosan must degrade the thermodynamic quality of the solvent and increase the degree of structuring and dynamic viscosity of concentrated polymer solutions. A monotonic decrease of the intrinsic viscosity and growth of Huggins constant in the studied range the ethanol ratio: water (curves 1 and 2 in Figure 1, respectively) confirm the conclusion about the deterioration of the thermodynamic quality of mixture acetic acid-ethanol solvent.

The dates of Figures 2 and 3 about the dynamic viscosity of 1–8% solutions of chitosan, shows the viscosity increase with the introduction of alcohol, more than have more maintenance of alcohol and chitosan. They also consistent with the known views on the impact of solvent quality on the properties of the solutions. However, the introduction of ethanol visually observed decrease in turbidity and increase the homogeneity of concentrated (6–8%) solutions chitosan in dilute acetic acid. The decrease of turbidity for 1% solutions is confirmed by calculation (curve 1 in Figure 4). The calculations also showed a significant reduction in the number of scattering particles (curve 2) and some increase in their average sizes (curve 3) with the introduction of ethanol, apparently due to dissolution micro and nanosized helices.

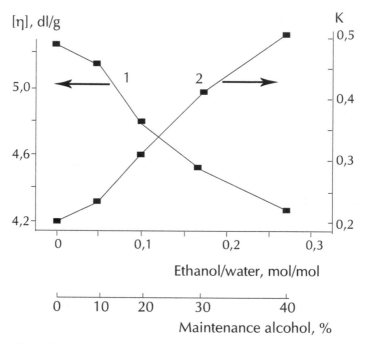

FIGURE 1   The influence of the solvent composition on the intrinsic viscosity (1) and Huggins constant (2).

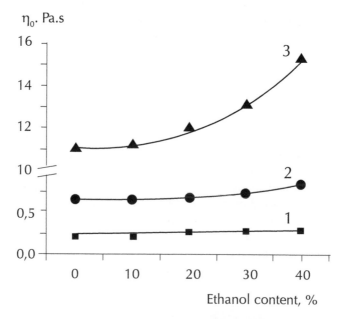

FIGURE 2   The dependence of the viscosity of 1 (1), 3 (2) and 8% (3) solutions of chitosan on the content of ethanol the increase of viscosity with the introduction.

All of this suggests a more complex effect of ethanol on the structural and rheological properties of the chitosan acetic acid ethanol–water systems. It was interesting to evaluate the influence of alcohol on the relative viscosity of solutions. According to our measurements the dynamic viscosity of the three component solvent water-alcohol acetic acid increases more than twice from 1.0–2.4 MPa c with the content of ethanol to 40%. According to [13], the viscosity water-ethanol mixtures with alcohol content of 42% have a maximum value of 2.8 MPa c that indicates a high degree of structuring of the solvent. As the chitosan solution relative viscosity (Figure 5, curves 1¢–3¢), reflecting the contribution of the polymer to the bulk viscosity decreases with the introduction of alcohol, thus increasing the viscosity of alcohol solutions of chitosan is due to a large extent the propertics of the solvent, structural associates whose in concentrated solutions of polymer formed macroclasters significantly increase the initial viscosity of the system.

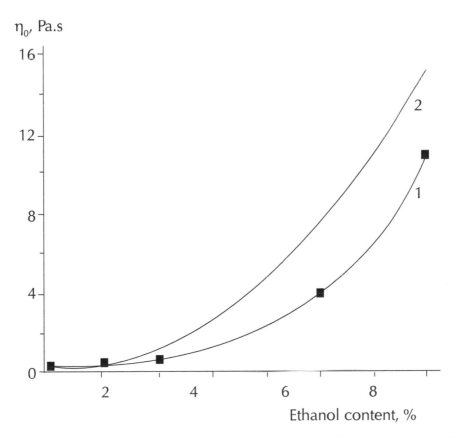

FIGURE 3 The concentration dependence of viscosity of chitosan solutions in the absence (1) and in the presence of ethanol 40% (2).

N. 10^{-9}, sm^{-3}

τ. 10^{1}, sm^{-1} r. 10^{-2}, nm

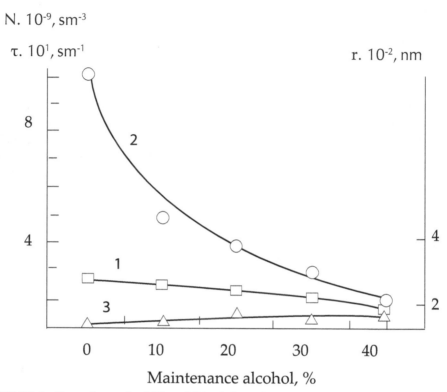

FIGURE 4 Dependence of turbidity of 1% solutions of chitosan (1), numbers (2) and particle size (3) on the content of ethanol.

The results of the temperature dependence of solutions viscosity and the calculation of the activation energy of viscous flow (Ea), as well as changes in the structuring of the index (n) showed similar values of these variables: in 8% solutions containing 40% alcohol and do not contain it, Ea = 40 and 36 kJ/mol, n = 0.75 and 0.80 (in the range of shear rate 10–50 s^{-1}), suggesting that there was no significant difference in the strength of chitosan's associates in solutions with ethanol and without ethanol. Therefore we can assume that the high viscosity of chitosan solutions containing alcohol is due to the increased degree of structuring of the solvent. Thus, the observed increase in viscosity of the spinning solutions of chitosan and reducing turbidity in adding them ethanol is not due to an increase in the size of the macromolecules associates in a thermodynamically poor solvent, but rather dissolution primary associates liquid micro and nanocristals, possibly due to heating of the mixture through energy released during the exothermic process of interaction of ethanol with water.

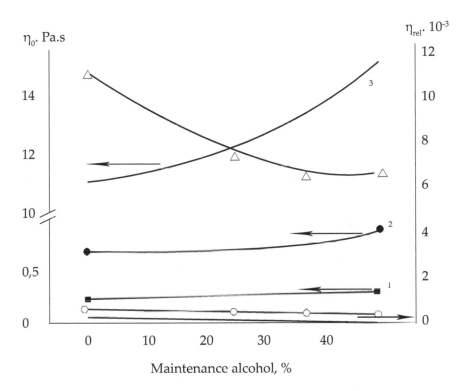

FIGURE 5 Dependence of viscosity (1–3) and the relative viscosity (1'-3) of chitosan solutions 1 (1,1¢), 3 (2,2¢) и 8% (3,3¢) on the ethanol content.

In favor of such explanation speak bigger turbidity of 1% chitosan solutions pre-pared by directly dissolving polymer in ethanol-containing solvents (20 and 40%). As well as short-term reversible decrease the viscosity of the system when adding the next portion of ethanol, visually observed as an increase in speed mixers, significantly exceeds that due to dilution of an equal amount of water.

Instability of viscous properties of chitosan solutions and the fall of viscosity during storage noted above. As shown in [8], the cause is degradation of macro-molecules, to warrant the subsequent destruction of the structure of solutions. Re-duced viscosity can slow by increasing of the acetic acid concentration, but since the reduction of viscosity spinning of polymer solutions facilitates their processing (transport, filtration), this process can be regarded as a kind of "maturation" solu-tions. It was of interest to characterize the influence of ethanol on the stability of viscous properties of chitosan solutions. According to our data (Figure 6), the introduction of alcohol accelerates the process of reducing the viscosity of solutions of chitosan.

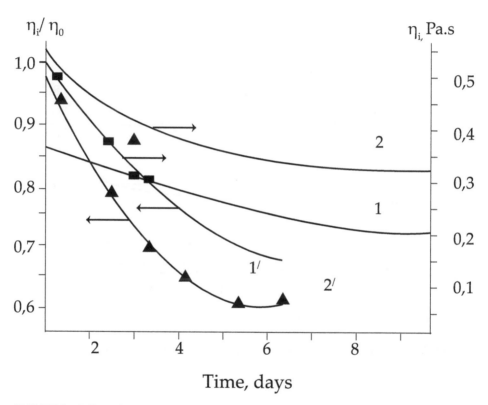

FIGURE 6 Effect of storage duration on the dynamic viscosity of 3% chitosan solutions in the presence of ethanol 40% (1,1') and in the absence of ethanol (2,2').

In this case the viscosity of ethanol-containing solutions remains at a slightly higher level compared with the viscosity of solutions without ethanol. Establish reasons for the lower stability of viscosity of water-ethanol solutions of chitosan and the rate of degradation of the polymer in them requires additional investigation.

Possibility and efficiency of use ethanol containing chitosan's solutions for preparation of products was shown for example in a formation chitosan covers on a viscose thread and creations fibrous chitosancontaining sorbent. Such composite corbent received by stretching an initial viscose thread through a chitosan solution and the calibrated aperture spinneret, deleting excess of a solution and subsequent evaporation of solvent.

Threads with the chitosan maintenance 10% have the expressed structure "core-shell". The deformational and strength properties of threads practically do not change up to the chitosan maintenance 20% (relative strength and elongation of the initial and modified threads are 16.8 against 16.5 sN/teks and 17 against 14%, accordingly). The cover put from chitosan's solution with ethanol have more uniformity, it is process evaporation passes with higher speed and has more stability (Figure 7).

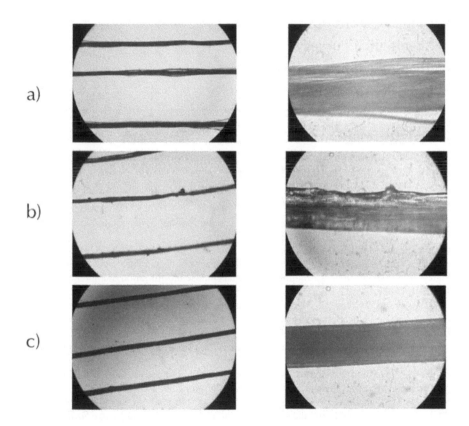

FIGURE 7   An initial viscose thread (a) a thread with a chitosan cover put from solution without ethanol (b) and a thread with a cover put from ethanolcontaining solution.

6.4   CONCLUSION

Thus, this study substantiates the possibility of introducing of ethanol into the spinning chitosan's solutions, because it does not lead to phase separation in solution, a significant increase in its initial viscosity and does not affect the viscosity of the 8% solutions with a gradient of shear rate over 10 s^{-1}, that is at the application of even a small mechanical impact, for example, when mixing and transportation solutions. In preliminary experiments showed that the presence of ethanol in solution significantly accelerates the process of curing chitosan films and membranes deposited on viscose filament increases the stability of the process of forming the envelope and received by the uniformity of the filament.

KEYWORDS

- **Acetic acid**
- **Chitosan solutions**
- **Rheological curves**
- **Rheological properties**
- **Viscosity of solvents**

ACKNOWLEDGEMENTS

This work was supported by the Ministries of Education and Science of the Russian Federation (the contract № 16.740.11.0059)

REFERENCES

1. Papkov, S. P. *Physicochemical Principles of Processing Polymer Solutions*. Moscow: Chemistry (1971).
2. Okihito, H. and Masatosi, H. *J. Chem. Soc. Jp. Ind. Chem. Sci.* A147–R12 (1973).
3. Sedelkin, V. M., Denisov, G. P., Surkov, A. N., Ramazaeva L. F., and Pachina O. V. (2007). Rheological properties of molding solutions for acetatcellulose ultrafiltration membranes. *Chemical Fibers*, **1**, 15–17.
4. Ma, O., Lavertu, M., Sun, J., Nguyen, S., Buschmann, M. D., Winnik, F. M., and Hoemann, C. D. Precise derivatization of structurally distinct chitosans with rhodamine B isothiocyanate. *Carbohydrate Polymers*, **72**, 616–624 (2008).
5. Hirano, S., Ohe, Y., and Ono, H. Selective N-acylation of chitosan. *Carbohydrate Polymers*, **2**, 315–320 (1979).
6. Hirano, S., Nagamura, K., Zhang, M., and Kim, S. K. Chitosan staple fibre and their chemical modification with some aldehydes. *Carbohydrate Polymers*, **39**, 293–298 (1999).
7. Sklyar, A. M., Gamzazade, A. I., Rogovina, L. Z. et al. The study of rheological properties of dilute and moderately concentrated solutions of chitosan. *Vysokomolekulyarnye Soedineniya*, **23**, 6, 1396 (1981).
8. Nud'ga, L. A., Petrov, V. A., Bochek, A. M. et al. Molecular and supramolecular transformations in solutions of chitosan and allylchitosan. *Polymer Science*, **39**, 259 (1997).
9. Mironov, A. V., Vikhoreva, G. A., Uspenskiy, S. A., and Kildeeva, N. R. Reasons for Unstable Viscous Properties of Chitosan Solutions in Acetic Acid. *Polymer Science*, **49**(1–2), 15 (2007).
10. Rinaudo, M., Pavlov, G., and Desbrieres, J. Influence of acetic acid concentration on the solubilization of chitosan. *Polymers*, **40**, 7029 (1999).
11. Vikhoreva, G. A., Rogovina, S. Z., Pchelko, O. M., and Galbraikh, L. S. The Phase State and Rheological Properties of Chitosan–Acetic Acid–Water System. *Polymer Science*, **43**(5–6), 166–170 (2001).

12. Klenin, V. I., Shchyogolev, S. Y., and Lavrushin, V. I. *Characteristic features of light scattering of dispersed systems*. Dr. S. Ya Frenkel (Ed.), Publisher, Saratov University, Saratov (1977).

13. Afanas'ev, V. N., Efremova, L. S., and Volkova, T. V. *Physicochemical Properties of Binary Solvents: Aqueous Systems*. Institute of Chemistry of nonaqueous solutions of RAS, Ivanovo (1988).

7 Updates on Polymer-Inorganic Materials

I. Yu. Yevchuk, O. I. Demchyna,
V. V. Kochubey, H. V. Romaniuk, Z. M. Koval',
G. E. Zaikov, and Yu. G.Medvedevskikh

CONTENTS

7.1 INTRODUCTION

Recently sol-gel synthesis of organic-inorganic hybrids has received an extensive attention. Sol-gel technique allows obtaining a lot of materials with a wide range of application. Nowadays researchers' attention is concentrated on the problem of obtaining of proton conductive materials. Such, materials can be used as polymeric electrolytes and proton conductive membranes in fuel cells, gas sensors, and solar cells [1, 2].

Polymers containing sulfonic groups are the most widespread proton conductive polymer materials. There are some methods of synthesis of such materials. At direct heterogeneous sulfonation of polymers by sulfonic groups the latter are disposed mainly on a surface; therefore, it is difficult to achieve the homogeneous structure of material. Destructive action of some sulfuring agents is the lack of sulfonation of solutions of polymers. Using another method polymerization of sulfonated monomers results in obtaining of polymers with high water uptake, what worsens mechanical properties of material? The subsequent cross-linking of such polymers is a high cost process.

Another method of synthesis of polymers containing sulfonic groups is condensation of polymers with sulfocompounds. The authors [3] have shown possibility of obtaining of films on the basis of products of compatible condensation of aliphatic polyamides, *n* – phenolsulfonic acid and formaldehyde in the medium of organic solvent. The authors [4] suggest preparing film material by condensation of dissolved poly(vinylidene fluoride), n – phenolsulfonic acid and formaldehyde. Method [5] of forming proton conductive membranes by chemical cross-linking of poly(vinyl alcohol) by glutaraldehyde with addition of poly(styrene sulfonic acid) is offered. However, these materials do not possess the sufficient level of proton conductivity.

The most commercially successful are membranes Nafion (Du Pont, USA). Nafion is perfluorosulfonic ionomer with high proton conductivity and chemical and mechanical stability [6]. However, limited operation temperature, low water uptake, and high cost are disadvantages of the perfluorosulfonic polymer.

Hence, alternative preparation methods for proton conductive materials are required. The perspective approach seems to be the usage of composite materials. The combination of organic polymers and ceramics promises new hybrid materials with high performance. They may be prepared by incorporation of nanoscale filler particles into polymer matrix. In [7] nanoparticles of titanium (IV) oxide were added into triazol-containing proton conductive membranes. Authors [8] improved conductivity of sulfonated poly(arylene etherketone) by incorporation of nanodisperse additives acid zirconium phosphate. In [9] it was suggested to introduce inorganic phosphosilicates into membrane material, and in [10] nanoparticles of silica functionalized by sulfogroups.

A simple method for obtaining an organic-inorganic composites is mixing of an organic polymer with alkoxide (e.g. tetraethxoysilane) followed by a sol-gel process involving hydrolysis and polycondensation of precursor. Such method provides new opportunities for preparing both inorganic and organic-inorganic two and multicomponent composites at a relatively low temperature. In [11] silico phosphate xerogels have been obtained on the basis of tetraethoxysilane and phosphoric acid and under the pressure of 5,000 kg/cm^2 electrolyte membranes with high proton conductivity ($10^{-3}10^{-2}$ Sm/cm) have been prepared. However, these membranes have low mechanical durability.

Therefore, it is reasonably to conduct sol-gel transformation in a matrix of soluble organic polymer or in monomer during simultaneous polymerization. By these methods it is possible to obtain hybrid organic-inorganic materials possessing high proton conducting and mechanical properties.

7.2 EXPERIMENTAL

For researches reagents were used as follows: Tetraethoxysilane $Si(OC_2H_5)_4$ ("EKOS-1", Russia), ethanol ("r.g."), orthophosphoric acid ("r.g."), and poly(vinylidene fluoride) (PVDF) M$_w$ 175000 (Aldrich).

Polymerizing mix "Discophot-1" consists of: Tetramethylene acrylate (TMDA) 62, epoxyacrylate oligomer 32, monofunctional vinyl monomer N-vinylpyrrolidone 5, and photoinitiator 2,2-dimethoxy-2-phenylacetophenone—1(% mass.). Such composition of polymerizing mix provided insignificant shrinkage at hardening.

Organic-inorganic composites were synthesized as follows: first, 10% wt PVDF solution was prepared by dissolution polymer powder in dimethylformamid at temperature of 40°C. Then, required quantities of sol-gel system (SGS) TEOS:C_2H_5OH:H_3PO_4:H_2O were added slowly into PVDF solution to obtain composites with different PVDF:TEOS ratio. Obtained systems were mixed using a magnetic mixer during 2hr at 40°C and then were cast on glass substrate for films forming.

Alternatively, organic-inorganic films were prepared *via* photoinitiated polymerization of diacrylate composition "Discophot-1" in the presence of above mentioned SGS. Kinetics of the process of photoinitiated polymerization of investigated systems was studied by laser interferometry. Photoinitiated curing of composition was conducted in a thin layer under cover glass at UV-irradiation (intensity of UV-irradiation was 14 W/m^2). The changes of intensity of interference picture during contraction of polymerizing composition were registered by a photodetector. Relative integral degree of monomer transformation or conversion P was calculated as

$$P = H_t / H_o \qquad (1)$$

where H_t is a layer contraction in the moment of time t, and H_o is a maximum achievable contraction, determined after the number of peaks on interferogram.

Measurements of proton conductivity of the samples were performed with the use of impedance spectrometer "autolab" (Ecochem, Holland) with the FRA program over the frequency range of $10 \cdot 10^5$ Hz. Samples were sandwiched between platinum electrodes by a diameter of 1 cm. A value $1/R_F$ was considered as a value of proton conductivity; R_F = a segment on an axis of real resistance in impedance hodograph [2]. Specific proton conductivity was determined after a formula:

$$\sigma = l/RS \qquad (2)$$

where R = sample's resisance, Ohm, l = sample's thickness, cm, and S = electrodes' area, cm^2.

Complex thermal analysis (thermogravimetric and differential thermal) of composite samples was conducted by means of Derivatograph Q-1500D (Paulik-Paulik-Erdey) under dynamic conditions in a temperature range of 20–400°C. The Heat of samples with mass of 200 mg was carried out in an air atmosphere at a heating rate of 5°C/min. Aluminum oxide was used as a reference.

7.3 DISCUSSION AND RESULTS

An operating temperature of fuel cells is required to be 120–130°C and more for providing an effective desorbtion of carbon oxide (CO), which is usually present in hydrogen fuel and "poisons" a platinum catalyst. Therefore, heat resistant polymers are used for making proton conductive membranes.

For our investigations we have chosen poly(vinylidene fluoride) since it is famous for its outstanding chemical and thermal stability as well as mechanical strength. Forming of inorganic structure took place *in situ* in polymeric matrix of PVDF using sol-gel technique. Sol-gel solution, consisting of TEOS, water, ethanol, and orthophosphoric acid (TEOS:C_2H_5OH:H_3PO_4:H_2O = 2,2:7,24:0,2:0,36 p. v.), was added into PVDF solution in dimethylformamid. As a result of, TEOS hydrolysis sol of polysiloxane particles appeared in polymeric matrix. Orthophosphoric acid served as a catalyst of

TEOS hydrolysis. Sol structuring at temperature of 40°C in constant temperature oven resulted in forming of organic-inorganic films.

Proton conductivity of nanocomposite films was determined by measuring of complex resistance impedance. Real and imaginary constituents of a vector of impedance allow evaluating conducting property of material. In Figure 1 one can see the dependence of real and imaginary constituents of impedance on current frequency and Nyquist plot of the cell Pt sample—Pt for the sample PVDF:TEOS = 30:70 (% w.) within a frequency range of $10–10^5$ Hz. Proton conductivity of investigated samples was about 10^{-4} Sm/cm.

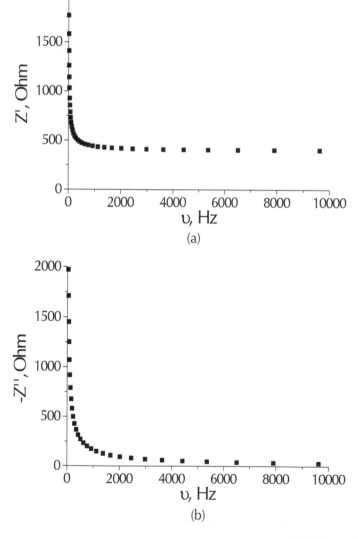

(a)

(b)

FIGURE 1   *(Continued)*

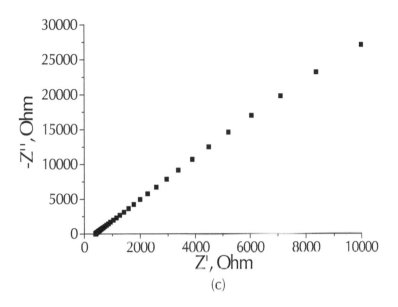

(c)

FIGURE 1   Dependence of real (a) and imaginary (b) constituents of impedance on frequency and Nyquist plot (c) for the sample PVDF:TEOS = 30:70 (% w.).

Ion conductivity in these composites is provided by inorganic component formed as a result of sol-gel process. In accordance with Grotthus concept transmission of protons takes place in water medium along channels due to the continuous exchange H_2O + H^+ = H_3O^+. Other mechanism assumes that protons pass by two ways: *via* diffusive transport of H_3O^+ ions and *via* rotation of proton containing groups [12]. Obviously, such groups are silanol and P-OH ones.

Thermal stability of prepared material was investigated by thermogravimetric and differential thermal analyses, the results of which are shown in Figure 2. Weight loss of the sample is observed at temperature range of 20–150°C. It is accompanied by appearance of endothermal effect in DTA curve. These effects may occur due to the presence of dimethylformamid remains. The second endothermal effect with maximum at 150°C in DTA curve can be observed at temperature range of 135–185°C. Probably, it is caused by the process of PVDF melting. Intensive weight loss of samples at temperatures higher than 385°C can be explained by the process by deep thermo oxidizing destruction of polymer. Hence, this composite material is thermally stable up to approximately 135°C, what is important for using it as proton conductive membranes.

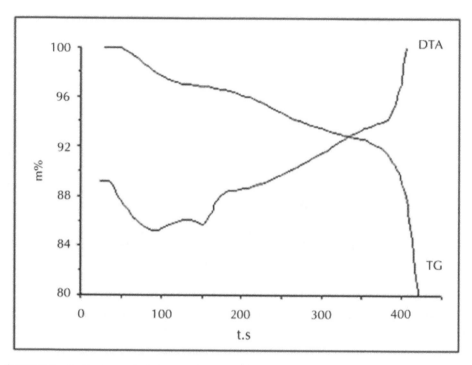

FIGURE 2   Derivatograph curves of composite membrane (PVDF:TEOS = 30:70% w.).

Photoinitiated polymerization of diacrylate polymerizing composition (PC) in the presence of SGS was the alternative way of synthesis of polymer-silica films. Kinetics of polymerization was studied depending on gelation time, on concentration of catalyst – orthophosphoric acid in SGS, as well as on mixture composition. The obtained results are presented in Figure 3–5 and in Tables 1–3.

As one can see in Figure 2–4, kinetic curves of polymerization in the presence of SGS have typical S-like shape. However, the rate of photoinitiated polymerization of diacrylate composition with increase of SGS content decreases as compared with initial polymerizing composition. Maximal rate of polymerization w_{max} at SGS content of 70% v. diminishes approximately by 2 times in comparion with w_{max} for initial composition, whereas, time of achievement of maximal rate increases by 2 times. Possible explanation of this fact may be that an inorganic constituent forms steric limitations for the process of polymerization of diacrylate monomer. An additional spatial network of nanoparticles of silica phase which appears as a result of sol-gel process leads to macroradicals decay and accordingly, to deceleration of polymerization process.

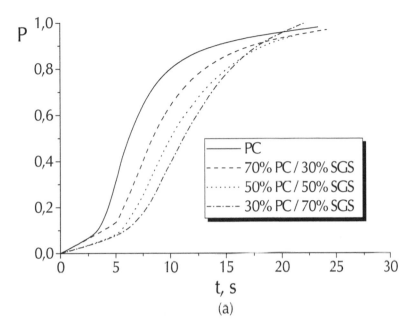

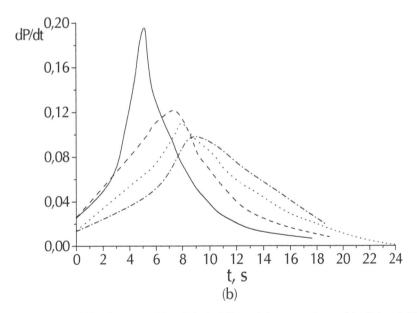

FIGURE 3 Integral kinetic curves (a) and their differential anamorphoses (b) of photoinitiated polymerization of PC-SGS depending on its composition.

TABLE 1   Kinetic parameters of the process of photoinitiated polymerization of PC-SGS depending on its composition.

№	PC - SGS, % v.	Time of w_{max} achiev. t, s	Conversion at w_{max}, P	Max. rate w_{max}, s^{-1}
1	100:0	5.2	0.34	0.203
2	70:30	7.3	0.38	0.124
3	50:50	9.1	0.41	0.108
4	30:70	10.4	0.43	0.102

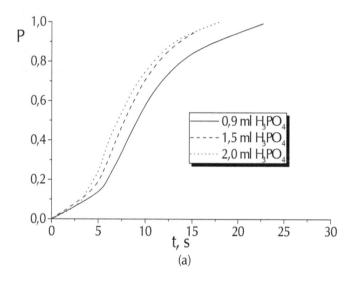

(a)

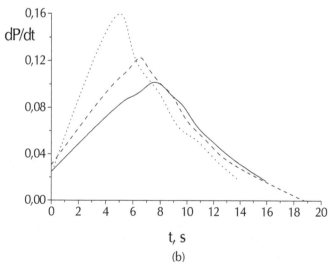

(b)

FIGURE 4   Integral kinetic curves (a) and their differential anamorphoses (b) of photoinitiated polymerization of PC:SGS = 50:50 (% v.) depending on catalyst concentration.

TABLE 2 Kinetic parameters of the process of photoinitiated polymerization of PC:SGS depending on catalyst concentration.

№	PC:SGS, % v.	H_3PO_4 in SGS, %v.	Time of w_{max} achiev., t, s	Conversion at w_{max}, P	Max.rate w_{max}, s^{-1}
1	50:50	9	7,4	0.33	0.10
2	50:50	15	6,6	0.31	0.13
3	50:50	20	5,4	0.29	0.19

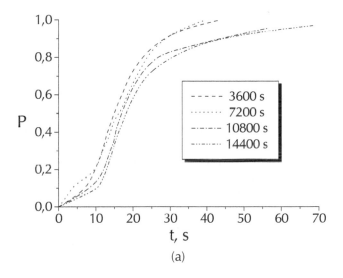

(a)

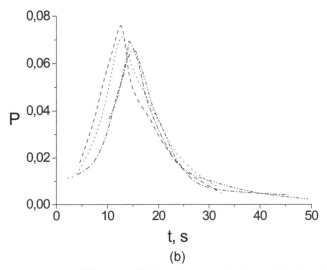

(b)

FIGURE 5 Integral (a) and differential (b) kinetic curves of photoinitiated polymerization of PC:SGS (9% v. H_3PO_4) = 50:50 (% v.) depending on gelation time.

TABLE 3   Kinetic parameters of the process of photoinitiated polymerization of PC:SGS (9% v. H_3PO_4) depending on gelation time.

№	PC:SGS, %v.	Time of gelation, S	Time of w_{max} achiev., t, s	Conversion at w_{max}, P	Max. rate, w_{max}, s^{-1}
1	50:50	3600	12,90	0.39	0.082
2	50:50	7200	13,10	0.35	0.075
3	50:50	108,00	14,85	0.34	0.071
4	50:50	144,00	15,08	0.33	0.067

At introducing into polymerizing composition SGSs with larger gelation times we also observe decreasing of maximal rate of the process for the same reasons.

At varying of orthophosphoric acid content in SGS from 9 to 20% v. maximal rate of polymerization increases almost by two times. We suggest the change of aggregation character of inorganic phase: At large concentration of catalyst the process of TEOS hydrolysis passes with high rate, it is therefore, possible to assume, that nanoparticles of silica phase are incorporated into polymer network, and aggregates, which could be "traps" for macroradicals do not form.

Proton conductivity of composites, obtained by this method, was measured by impedance spectrometry over frequency range of 10^{10^5} Hz. In Figure 6 one can see the dependence of real and imaginary impedance constituents on current frequency as well as Nyquist plot for the sample PC:SGS = 50:50 (% v.) at catalyst concentration in SGS of 20% v.

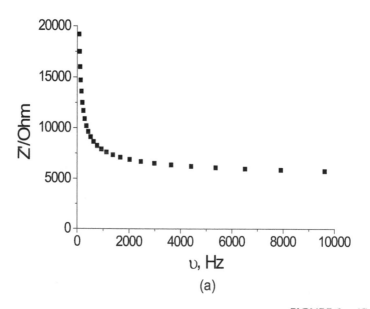

(a)

FIGURE 6   *(Continued)*

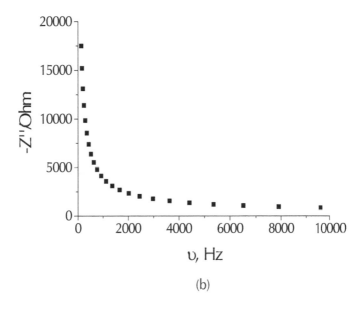

(b)

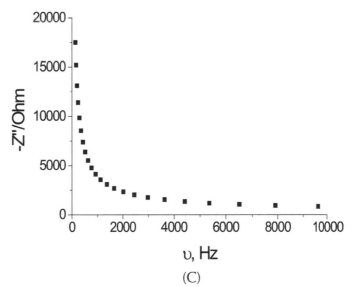

(C)

FIGURE 6   Dependence of real (a) and imaginary (b) constituents of impedance on frequency and Nyquist plot of impedance (c) for the sample PC:SGS = 50:50 (%v.).

Proton conductivity of composites, obtained by photoinitiated polymerization, was 10^{-6} Sm/cm. This value is by two orders smaller than the values of proton conductivity of nanocomposites on the basis of PVDF. It was also found to be depended on catalyst concentration (Table 4).

TABLE 4　Proton conductivity of the samples PC:SGS.

PC:SGS = 50:50 (% v.)	
H_3PO_4 in SGS, % v.	σ, Sm/cm
9	1.72×10^{-6}
20	4.13×10^{-6}
30	9.41×10^{-6}

7.4　CONCLUSION

Polymer-silica nanocomposites were synthesized by two methods—by forming of silica phase as a result of sol-gel process *in situ* in polymeric matrix of PVDF and during photoinitiated polymerization of composition on the basis of diacrylate monomer TMDA. Organic-inorganic materials, obtained by the first method, possess proton conductivity of 10^{-4} Sm/cm and can be used at temperatures up to 135°C.

KEYWORDS

- **Nafion**
- **Organic-inorganic composites**
- **Polymer-silica nanocomposites**
- **Proton conductive materials**
- **Sol-gel system**
- **Sol-gel technique**
- **Sulfonic groups**

REFERENCES

1. Maltseva, T. V. Inorganic proton conductive nanomaterials: Outlook for application in membrane fuel cells. *Nanosystems, nanomaterials, and nanotechnologies*, **2**(3), 875–894 (2004).
2. Dobrovolsky, Yu. A., Pisareva, A. V., and Leonova, L. S. et al. New proton conductive membranes for fuel cells and gas sensors. *Alternative energy and ecology*, **12**(20), 3641 (2004).
3. Kobelchuk, Yu. M., Chervakov, O. V., and Gerasymenko, K. O. et al. Synthesis of sulfonated derivative polyamids and film materials on their basis. *Voprosy khimiyi I chim. tekhnologiyi*, (1), 7883 (2008).
4. Chu, P. J., Wu, C. S., and Chen, J. Y. PVDF-HFP/P-Sulfonate-phenoline DMFC membrane by in situ synthesis. Proceedings of 2003 Fuel Cell Seminar: Book abstract. Miami Beach, Florida. N 37, pp. 474–477.
5. Stadny, I. A., Konovalova, V. V., and Yevdokymenko, V. O. et al. Proton conductive membranes based on polyvinyl alcohol and polystyrenesulfonic acid. *Chemical sciences*, Magisterium, **33**, 38 (2008).
6. Shi, Z. and Holdcroft, S. Synthesis of Block Copolymers Possessing Fluoropolymer and Non-Fluoropolymer Segments by Radical Polymerization. *Macromolecules*, **37**(6), 2084–2089 (2004).
7. Danyliv, O. I., Konovalova, V. V., and Burban, A. F. Developoment of method of triazol-containing proton conductive membranes formation. *Scientific proceedings*. Chemical sciences and technologies. – NU "Kyiv-Mohyla Academy", Kiev. "Pulsary", **92**, 1218 (2009).

8. Fomenkov, A. I., Pinus, I. Yu., and Peregudov, A. S. et al. Proton conductivity of poly(arylene ether ketones) with different sulfonation degrees: Improvement *via* incorporation of nanodisperse zirconium acid phosphate. *Vysokomol. soed.* **49**(7), 1299–1305 (2007).
9. Yonggang, Jin, Joao, C., and Diniz, da Costa. et al. Proton conductive composite membrane of phosphosilicate and polyvinyl alcohol. *Solid State Ionics*, **178**, 937–942 (2007).
10. Martinelli, A., Matic, A., and Jacobsson, P. et al. Structural analysis of PVA-based proton conducting membranes. *Solid State Ionics,* **177**, 2431–2435 (2006).
11. Shilov, V. V., Shilova, O. A., and Yefimova, L. N. et al. *Perspective materials*, (3), 3137 (2003).
12. Komarov, P. V., Veselov, I. N., and Khalatur, P. G. Nanoscale morphology in ionic membranes on the basis of sulfonated aromatic poly(ether ketones): Mesoscopic simulation. *Vysokomol. Soed*, **52A**(2), 279–297 (2010).

8 A Note on Gradient Refractive Index

L. Nadareishvili, Z. Wardosanidze,
N. Lekishvili, N. Topuridze, S. Kubica,
and G. Zaikov

CONTENTS

8.1 INTRODUCTION

Creation and investigation of materials with the gradient of properties is considered one of the main directions of polymer science for the 21st century. In this direction, the essential success was achieved in the 1970s of the previous century. The methods of obtaining materials with inhomogeneous distributions of properties were developed. The heterogeneity of composition conditions the gradient of refraction. In such an area a trajectory of the light beam is constantly curvilinear along the free path length, which causes deviation of light beam, and during proper distribution of refraction index focusing of beams, too. In the 1970s, the cylinder elements were fabricated from polymers with radial parabolic distribution of composition, having the ability of transmission of light and focusing. These elements received the name Selfoc (Self focusing).

By introducing a new parameter in optics—refraction index gradient, — new conditions of theoretical and experimental investigations have been created. The fundament was laid both for wide possibilities for creation of radical improvement of existing optical devices and new optical devices, fabrication of which on the basis of traditional optical materials have been excluded principally. In scientific literature for indication of refraction gradient elements (materials and areas) there has been established a term GRIN (Gradient Refractive Index), and for corresponding field of science GRIN optics.

For today GRIN optics is the independent perspective direction. The GRIN elements can independently form and translate an image without additional means. They are widely used in optics and optoelectronics of various destinations (Fiber-optical lines of communications, the agreement elements, endoscopies systems of small sizes, copier, lenses of low chromatic aberrations, the focusing elements of video recording laser systems, flat lenses, etc). In each technologically developed country the intensive investigations in this field are being carried out. The special laboratories and research centers have been created. We have received some interesting results in this sphere.

The world practice of development of GRIN optics indicates that the attention of the researchers is concentrated only on the refraction index. At the same time, experts of GRIN optics consider similarly perspective the materials having other optical properties, in particular, materials of gradient birefringence, too. The first polymer materials (films) having such optical properties were obtained by us. We also contributed the acronym for the materials with gradient birefringence (elements and areas) GB (Gradient Birefringence) — material (element and area).

By introducing a new parameter in optics—Gradient Birefringence—a fundament of a new direction of optics was laid; It resulted in widening the notion of "Gradient Optics". Nowadays, the gradient optics covers two independent directions—GRIN optics and GB optics. Both these directions are generally strategic in development of polymeric science of gradient material science.

The ways of formation of GRIN and GB elements are different. The GRIN elements are received as a result of chemical transformations. Many methods have been elaborated, by means of which the known chemical transformations are realized in monomer (polymeric) systems in gradient regime (in selected directions and proper velocity).

Formation of GB elements is based on ability of great deformations characteristic to polymers. At the same time while realizing great deformations polymer passes on to specific, so called oriented state. Its essence is expressed in the macromolecule chains (as usual on separate sections) of polymeric body that have privileges locations in the whole polymeric body in any direction which is called an orientation axis.

The most widely spread of polymeric orientation is monoaxial orientation stretching. As a result of the orientation the physical and chemical properties of the whole number of exploitation properties of polymers are changed essentially. That is why the studying of orientation processes which is an integral part of polymers structure study (molecular and super molecular) is one of the main tasks of polymer science.

With allowance of property gradient, as a strategic mark of development of polymeric science, the oriented state of polymers acquires a new essence, which can be qualified as gradiently oriented state.

We suppose that introduction of new structural characterization of polymers as a new physical characteristic of polymeric nature essentially broadens general problems of scientific research of polymers. It principally increases possibilities of regulation of mechanical, thermal, electric, optical, and other properties, giving an impetus to creation of new scientific directions as it has already happened on the example of GB optics. The essential characteristic of this position except gradient is the angle between structural orientation and the direction of the gradient. We can regulate it in the interval $90° \geq \alpha \geq 0°$.

For creation of gradiently oriented state the form of isotropic polymer sample (i.e., the form of clamps and interlocation) has been selected so that it is provided with preliminarily established gradient of relative lengthening in the sample. By means of variation of other parameters (temperature, value, velocity of deformations, etc.) it is possible to regulate gradiently oriented state of polymers. Characterization of this state is affected by means of observation on polarized light through the polarized microscope and studying of birefringence. Thus, the birefringence is a testing characteristic of gradiently oriented state and at the same time it is itself a purposeful property of the material.

Above the glazing temperature during the monoaxial linear polymer the polymer acquires symmetry of monoaxial crystal, the optical axis of which coincides with the stretching direction. The birefringence originated this time is functions of relative lengthening:

$$\Delta n = n_1 - n_2 = \gamma \lambda \tag{1}$$

There n_1 is an ordinary ray; n_2 = extraordinary ray; γ = optical coefficient of deformation; λ = relative lengthen. The Equation (1) is valid during the identity of all other parameters determined by the process mode.

We have elaborated out several versions of the equipment for creation of preliminarily established inhomogeneous mechanical field, as a result of cohesion in a polymeric body an established gradient of relative lengthening is formed, and consequently, so is the preliminarily established gradient birefringence.

For one series of the equipment, GB effect is achieved by the fact that the clamps allocated on one plane, in which a polymer film/plate of trapezoid shape rotates in interopposite direction around the parallel axis, the rims (edges) of clamps create φ angle at the rotation axis. In this case the distribution of lengthens (Δl) on the polymeric sample length (h) is expressed by the equation:

$$\Delta l = 2x.tg\varphi(1 - \cos\alpha) \tag{2}$$

There x is the length of the sample in a given point ($x = 0; x = h$); α is the angle of rotation of clamps.

In this work we consider some regularities of formation of the gradiently oriented state, which is connected with some properties of constructive determinations of corresponding apparatus. We investigate some optical properties of gradiently oriented polymers. There are discussed possible spheres of their application.

8.2   EXPERIMENTAL

Experiment was carried out on PVS films (thickness is 80100 mk).Tension of the film was made on device apparatus of specific construction (T = 358K, velocity of tension = 20 mm/min.). For creation of non-homogeneous mechanical field we use clamps, having various configurations.

We studied optical properties (gradient of anisotropy) on polarization micropho-tometer.

8.3   DERIVATION OF THE BASIS RELATIONSHIPS

Distribution of the relative lengthening is determined by profile of clamps $f_1(x)$ and $f_2(x)$, where independent variable x changes in the $[x_1, x_2]$ interval. $f_1(x)$ undergoes parallel displacing across the OY axis and $f_2(x)$ is fixed function. Let us designate the relative lengthening across the OY axis as $\Phi(x)$ and the value of parallel transfer as K, then for the distribution of the relative displacing we have:

$$\Phi(x) = \frac{K}{f_1(x) - f_2(x)} \qquad (3)$$

We can choose $\Phi(x)$ from the different class of function (linear, parabolic, hyperbolic, sinusoidal, etc.), take into account isochromic aspect of GB element. From the Equation (3) we have:

$$f_1(x) = K/\Phi(x) + f_2(x) \qquad (4)$$

Let us consider different cases of $\Phi(x)$ function:

(1) Linear function $\Phi(x) = ax + b$, then $f_1(x) = K/(ax + b) + f_2(x)$ $\qquad (5)$

when a = 1, b = 0, and the profile of the first clamp is linear, that is $f_1(x) = c$, then

$$f_1(x) = K/x + C \qquad (6)$$

Thus, the profile of the second clamp is hyperbola.

(2) Parabolic function $\Phi(x) = ax^2 + bx + c$, then: $f_1(x) = K/(ax^2 + bx + c) + C$ $\qquad (7)$

when a =1, b = 0, and c = 0,

$$f_1(x) = K/x^2 + C \qquad (8)$$

Thus, the profile of the second clamp is also hyperbola.

(3) Hyperbolic function $\Phi(x) = a/x$, then: $f_1(x) = Kx/a + C$ (9)

In this case the form of the second clamp is linear.

(4) Sinusoidal function $\Phi(x) = \sin x$, then: $f_1(x) = K/\sin x + C$ (10)

In this case the form of the second form is so complicated, that its realization is available only in the definitely, strictly limited conditions.

Let us assume that the initial length of the sample in the x point is more than distance between two clamps, that is $\Delta x > [f_1(x) - f_2(x)]$

Then for the relative lengthen we have: $\Phi(x) = K/(f_1(x) - f_2(x) + \Delta x)$ (11)

In this case the profiles of the clamps changes considerably:

(a) when $\Phi(x) = ax + b$ (linear function), then $f_1(x) = K/(ax + b) + f_2(x) - \Delta x$ (12)

Value of Δx is preliminarily chosen and always is linear function [i. e. $\Delta(x) = \alpha x + \beta$].

when $f_1(x) = C$, then $f_1(x) = K/(ax + b) + C - (\alpha x + \beta) = K/(ax + b) - \alpha x + \gamma$ (13)

where $\gamma = C - \beta$; when $a - 1, b - 0$, then: $f_1(x) = K/(x - \alpha x) + \gamma$ (14)

So, the profile of the second clamp is the linear combination of the hyperbola and line.
(b) When $\Phi(x) = ax^2 + bx + c$ (parabolic function), then:

$$f_1(x) = K/(ax^2 + bx + c) + C - (\alpha x + \beta)$$ (15)

In this case the profile of the second clamp is complicated function too.
(c) When $\Phi(x) = a/x$ (hyperbolic function), then:

$$f_1(x) = K/(a/x + c) - (\alpha x + \beta) = (K/a)x + c - \alpha x - \beta$$ (16)

So, the profile of the second clamp in this case is linear.

The obtained results give us a chance to improve possibilities of existing optical devices. The profile of the clamps with all those factors, which have influence on the oriented state of polymers (molecular and supermolecular structure, medium molecular mass and distribution of molecular mass, existence of ingredients and plasticators, velocity of tensile deformation, values of relative deformation, temperature, environment, and scale factor) define optical properties of GB elements.

We represent the experimental results to illustrate some specific cases of formation of gradiently oriented state (creation of GB elements).

8.4   PARALLEL CLAMPS

In this case the value of the relative displacement along of all the film perpendicularly to the direction of tension is constant. In according to this, there will be no gradient of birefringence. However, in polaroscope (crossed nicols, 6 times tension; the initial length of the film is 6 cm) sharp isochromes are seen on free ends of the film on the width of 1 cm. So, at the edge of the film the gradient of birefringence is in direction, perpendicular to the tension. There is a homogeneous region in the middle of the sample, where the birefringence (Δn) is constant. In figure 1(a) we have the microphotogram of film for the wavelength $\lambda = 630$ nm that is obtained by means of photoelectric polarization microphotometer (crossed nicols). Figure 1(a) corresponds to the picture observed in polariscope. There was shown that in free region of given wavelength two isochromic bands are received, maximum of which correspond to the half-wave regions $(2n + 1)\lambda/2$. Consequently, theoretically, a gradient will not take place when tension is parallel but experimentally we observe that the gradient is received. The reason of this may be free edges of the film. Absolute lengthening at the edges is greater, than in the middle. Such effect takes place in all cases of tension.

8.4.1   Clamps Directed by 45° Angles

In this case $f_1(x) = (\sqrt{2}/2)x$ and $f_2(x) = (\sqrt{2}/2)x$. Then the equation is transformed into:

$$\Phi(x) = K/x\sqrt{2} \qquad (17)$$

This is an equation of hyperbola. When we use such clamps for parallel tension of the film, we receive an oriented polymer film, where regulation of distribution of birefringence gradient perpendicularly to the tension direction is hyperbolic. Indeed, the regulation of distribution in perpendicular to the tension direction is hyperbolic. In Figure 1(b) we give microphotograms of these samples. Here the distances between bands corresponding to $(2n + 1)\lambda/2$ are nearly the hyperbolic distribution.

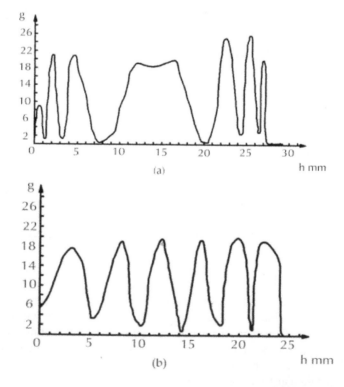

FIGURE 1   *(Continued)*

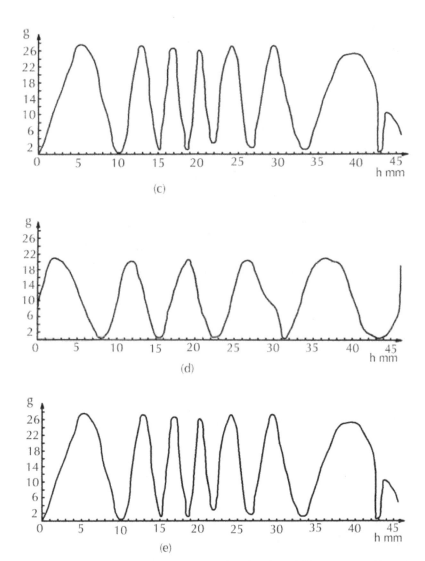

FIGURE 1   Dependence of transparency I of gradiently oriented film on h–coordinate (crossed Nichols): (a) Parallel clamps, the film of rectangle form, tension is perpendicular to the clamps, (b) Clamps allocated to the angle of 45° towards each other. Trapezium form Polymer film. The tension occurs in parallel to the base of trapezium, (c and d) Parallel clamps. Trapezium form Polymer film. Direction of tension is parallel to the attitude of trapezium. The length of one of free edges of the film (l2) is more than the distance between clamps l1, and (e) Parallel clamps. Trapezium form Polymer film. Direction of tension is parallel to the attitude of trapezium. Relation l2 / the distance between two clamps are more for one film (case c) than that for another (case d).

Parallel clamps. The length of one of free edges of the film is more than a distance between two clamps (Figure 2).

In this case, according to Equation (9), the distribution of the relative lengthens may be various. The tension is carried out until distance between two clamps equals to l_2. In Figures 1(c) and 1(d) we have micro photogram and we see that the distances between bands corresponding to $(2n + 1)\lambda/2$ are practically equal.

At the same time relation l_2 the distance between two clamps is more for one film (Figure 1(c)) than that for another (Figure 1(d)).

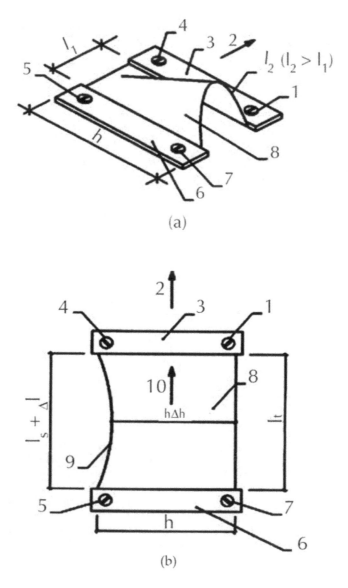

(a)

(b)

FIGURE 2   Gradient tension of polymer film 1,4,5, and 7 –screws; 2 –direction of tension; 3 and 6 –clamps; 8 -polymer film: (a) before tension; (b) after tension.

8.4.2 Parallel Clamps: Isosceles Trapezium and Tension Across the Attitude of Trapezium.

In this case isochromes are disposed near the big base of trapezium and parallel to it. In Figure 1(e) there is given corresponding microphotogram. In all cases described previously (cases 1, 2, and 3) direction of tension, disposition of isochromes, and gradient to each other is described by the scheme on Figure 3(a). This case is in contrast to other ones. Here disposition of these characteristics is different and is described by the scheme on Figure 3(b).

8.4.2.1 *Polarized Compensators*

Generally, the main principle of behavior of compensators is based on operation of exclusion of any optical parameter. For example, in case of ordinary isotropic phase compensators the thickness and refraction index ($n \cdot d$) of optical element have to provide with concrete phase shifting ($k.\lambda,...\infty$). Such compensators are used in interferometer tasks. In case of anisotropy compensators the notion of phase shift means not the absolute phase shifting for a given wavelength but rather the relative phase shifting ($\Delta n \cdot d$) between usual and unusual beams (waves).That is why that the accuracy of optical measurements have been increased considerably by using the anisotropy compensators. In spite of the aforementioned, the area of practical use of both the isotropic and anisotropy compensators is limited, so as practically they fulfill the function of a reference. In comparison to them, the GB compensator is not a reference optical element but it provides with both the concrete phase shifts in the whole section of visible spectrum and any phase shift for the given wavelength.

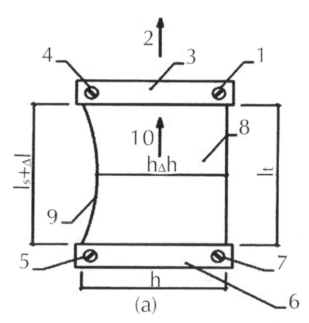

FIGURE 3   *(Continued)*

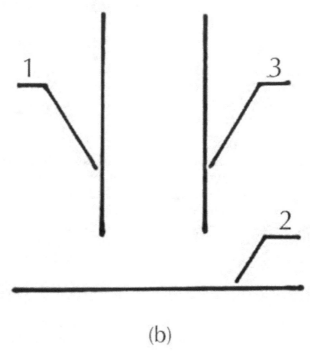

(b)

FIGURE 3   Direction of tension, disposition of isochromes, and gradient in gradiently oriented films: (1) Direction of tension; (2) Disposition of isochromes; h (3) Gradient: (a) Cases 1, 2, and 3; (b) Case 4. (4) Some spheres of application of GB elements

8.4.2.2   Polarized Holography and Photonics

Both in holography and photonics it has become necessary to envisage the polarized characteristics of the light. Generally in this case it is important only the intensity gradation was the reason why dynamic range of the process was sharply limited. In case of fixation of polarization it is already decisive not the intensity gradation (which is limited) but the fixation of practically boundless (infinite) versions of polarization. As a consequence in a given case it becomes necessary to provide with complex space polarized modulation of light. One of the ways for solution of this task is to use the GB elements having the complicated structure which gives possibility to realize the formation of light waves having the specific polarized characteristics.

A special importance is given to applying GB elements in photonics (in photo-chemistry), so as for today the investigations are being carried out very intensively for studying both the liner and nonlinear Veigert effects 1. This time the GB elements will provide with the accurate and simple relation between the linear and nonlinear Veigert effects. In particular, the calibrated (divided) GB elements give possibility to realize the radiation simultaneously by means of all kinds of polarization. This time it is very important that in corresponding sections of polarization the power exposition is absolutely similar, which is automatically realized. And this is particularly important, since it excludes the necessity of labor intensive and less accurate photometric measurements. In photonics the GB elements will fulfill the function a definite standard polarized modulators.

8.4.2.3   Interference Polarized GB Monochromator

The issue of miniaturization of spectral devices is very actually. It is clear that there are already created miniature monochromators in the form of gradient multilayer filters. The indicated filters are distinguished by the fact that the thickness of separate dielectric layers of multilayer systems is constantly changed along the filter which ensures a maximum transparency in a red section of spectrum on one its part, and in the violet section on the other side 2. The lack of such monochromators is the low capability of spectral distinction (>10A°). It is possible to improve the capability of spectral distinction if the interference and polarization filter is created under the similar principles. At the same time in such a monochromator it will be used not only the multilayer structure of alternative thickness but the additional gradient anisotropic structure (GB structure), the multiple of the wave length of which will be agreed compared ($\Delta n \cdot d = n \cdot \lambda$) with the maxims of transparency spectrum of the monochromator. The capability of distinguishing of such a monochromator is of 0.1A° order. The sizes of a monochromator may be from 5.1 mm^2 to 50–100 mm^2.

8.4.2.4   Luminescence Analysis

It seems very interesting the application of anisotropic films in luminescence analysis. In particular, as it is known in oriented organic films as in matrix. Radiation of installed luminescence paints is partially polarized. Besides, this time the degree luminescence (fluorescence) polarization mainly depends on the structure of luminescence material molecule itself and the quality of matrix orientation. In case if the matrix orientation changes along the film, we have a gradient anisotropic matrix, the luminescence polarization quality is characterized too by definite distribution, by the gradient to the same direction. If the degree of luminescence polarization, let us say along the X-axis, is expressed by $P_1(x)$, and the degree of matrix polarization or distribution of anisotropy by $P_m(x)$ or $\Delta n(x)$,

than the relative value: $K = P_1(x)/P_m(x)$ or $K = P_1(x)/\Delta n(x)$ (18)

In totally with luminescence intensity and spectral characteristics unambiguously should determine the definite features of given luminescence materials. This approach has definite properties as well for possible broadening of investigation of laser effects.

For development of GB optics (similarly as of classical optics) processing/development is a universal method which allows us to realize GB structures of concrete functional destination by means of creation of various technological equipment and controlling of their technical parameters. In case of simple one-dimensional compensators the main determining technological parameters can be the relative lengthens, profile and scale of clamps. In case of relatively complex two-dimensional compensators to this three main parameters are added the coordinated parameters which mean the simultaneous orientation to the orthogonal or generally to any other direction, too.

If we add to these materials a space modulation of temperature field as well, for example gradient heating the obtained GB element configuration can be changed within the wide ranges according to their destination.

KEYWORDS

- **Birefringence**
- **Gradient Refractive Index**
- **GRIN optics**
- **Parallel clamps**
- **Polymer materials**

REFERENCES

1. Kozlov, G. V., Shustov, G. B., and Zaikov, G. E. *J. Appl. Polymer Sci.*, **111**(7), 3026–3030 (2009).
2. Baranov, V. G., Frenkel, S. Ya., and Brestkin, Yu. V. *Doklady AN SSSR*, **290**(2), 369–372 (1986).
3. Vilgis, T. A. *Phys. Rev. A*, **36**(3), 1506-1508 (1987).
4. Rammal, R. and Toulouse, G. *J. Phys. Lett. (Paris)*, **44**(1), L13–22 (1983).
5. Klymko, P. W. and Kopelman, R. *J. Phys. Chem.*, **87** (23), 4565–4567 (1983).
6. Kopelman, R., Klymko, P. W., Newhouse, J. S., and Anacker, L.W. *Phys. Rev. B*, **29**(6), 3747–3748 (1984).
7. Nagaeva, D. A. *Dussert. ... kand khim nauk*, MkhTI, Moscow, p. 161 (1989).
8. Kozlov, G. V., Beeva, D. A., and Mikitaev, A. K. *Novoe v Polimerakh i Polimernykh Compositakh*, (2011).
9. Novikov, V. U. and Kozlov, G. V. *Uspekhi Khimii*, **69**(4), 378–399 (2000).
10. Alexander, S. and Orbach, R. *J. Phys. Lett.* (Paris), **43**(17), L625–631 (1982).
11. Askadskii, A. A. *The Physics-Chemistry of Polyarylates*. Khimiya, Moscow, p. 216 (1968).
12. Korshak, V. V., Pavlova, S. S. A., Timofeeva, G. I., Kroyan, S. A., Krongauz, E. S., Travnikova, A. P., Raubah, H., Shultz, G., and Gnauk, R. *Vysokomolek. Soed. A*, **24**(9), 1868–1876 (1984).
13. Budtov, V. P. *Physical Chemistry of Polymer Solutions*. Khimiya, Sankt-Peterburg, p. 384 (1992).
14. Family, F. *J. Stat. Phys.*, **36**(5/6), 881-896 (1984).
15. Kozlov, G. V., Mikitaev, A. K., and Zaikov, G. E. *Polymer Research J.*, **2**(4), 381–388 (2008).

9 IR Study of Silanol Modification of Ethylene Copolymers

S. N. Rusanova, O. V. Stoyanov, S. Ju. Sofina, and G. E. Zaikov

CONTENTS

9.1 INTRODUCTION

The chemical interaction of organosilicon compounds and copolymers of ethylene with vinyl acetate and copolymers of ethylene with vinyl acetate and maleic anhydride was studied by IR spectroscopy absorption and Attenuated Total Reflection (ATR). An enrichment of the surface layers of polymers by siloxane phase was found, that may be useful in the design of the chemical structure of adhesive materials for different purposes.

It is possible to extend the sphere of application of ethylene copolymers by the modification of the original polymer or by the development of composite materials on their basis. Traditionally, the compositions formation is effective in the presence of additives, interacting with the polymer during the processing. It results in the regulation of the material properties. One of the methods of chemical modification is the introduction of organosilicon compounds into polyolefins.

Earlier [1, 2], we studied the modification of ethylene copolymers with vinyl acetate by ethyl silicates containing various amounts of ester groups. Similar studies were conducted by Bounor-Legaré et al [3], so it was interesting to study changes in the structure and properties of ethylene copolymers with vinyl acetate containing other functional groups such as anhydride, to identify general trends and specific differences during the modification of ethylene copolymers, containing different reactive segments, by limiting alkoxysilane.

9.2 EXPERIMENTAL

9.2.1 Materials

Copolymers of ethylene vinyl acetate (EVA) grades Evatane 2020 and Evatane 2805 (Arkema) and Sevilen 11306–075 brand of JSC "Sevilen" (TU 6-05-1636-97); copolymer of ethylene vinyl acetate and maleic anhydride (EVAMA) grades Orevac 9305 and Orevac 9707 (Arkema) were used as the objects of the study. The main characteristics of the polymers are listed in the Table 1.

Ethylsilicate ETS-32 (ETS) (TU 6-02-895-78) being a mixture of tetraethoxysilane with geksaethoxydisiloxane with a small admixture of ethanol and oktaethoxytrisiloxane was used as a modifier. The silicon content in terms of silicon dioxide is 30-34%, tetraethoxysilane—50-65%. Its density is 1.062 kg/m^3 and viscosity is 1.6 cP. Ethylsilicate is manufactured by Production JSC "Khimprom" of Novocheboksarsk.

TABLE 1   Main characteristics of the polymers.

Characteristics	sevilen 11306-075	Evatane 2020	Evatane 2805	Orevac 9305	Orevac 9307
Symbol	EVA 14	EVA 20	EVA 27	EVAMA 26	EVAMA 13
The content of vinyl acetate, %	14	19-21	27-29	26–30	12-14
Melt Flow Rate, g/10 min, 125°C	0,85	2,23	0,53	14,06	1,10
Density, kg/m^3	0,935	0,936	0,945	0,963	0,924
Melting temperature (max), °C	97	82	72	67	90
Tensile strength, MPa	18,15	14,28	17,40	6,72	18,88
Breaking elongation, %	650	660	830	800	670

9.2.2 Samples Obtaining

Reactive blending of polymers with ethylsilicate was carried out on the laboratory rolls at a rotational rolls speed of 12.5 m/min and at a friction of 1:1.2 during 10 min in the range of 100–120°C. The content of the modifier was varied in the range of 0–10 mass %. The samples for investigations were prepared by the direct pressing in the restrictive frameworks. Pressing regime is under the temperature of 160°C and the unit pressure of 15 MPa; the time of preheating, injection boost time and the time of cooling is 1 min. for each 1 mm of the sample thickness. After the pressing all the compositions were subjected to aging at a room temperature for 24 hr.

Since, the rolling of two-component systems as the formation of products of chemical interaction between the components and the simple mechanical mixtures is possible. That is why a purification of the modified polymers was carried out by fivefold reprecipitation under cold conditions by the ethanol from the solution in CCl_4. The samples for Infrared (IR) spectroscopy absorption were prepared by watering from solution in carbon tetrachloride on a substrate of KBr. The film samples with the thickness of 0.07-0.12 mm for IR spectroscopy ATR were prepared by the direct pressing without restrictive frameworks on fluoroplastic plates.

9.2.3 Research Method

The IR spectra were registered by the infrared Fourier spectrometer "Spectrum BXII" of Perkin Company by absorption spectroscopy in the range of 450-4000 cm^{-1} and with the method of ATR on the ZnSe crystal in the range of 650-4400 cm^{-1} with the subsequent transformation by Kubelka-Munk. All spectra were normalized according to an internal standard and the intensity of the band of 720 cm^{-1} related to the deformation vibrations of CH_2 groups of the main chain not involved in the chemical reaction was assumed as the internal standard [1, 4]. The original spectra in the coordinates of the optical density wave number were processed using the software package ACD/SpecManager (ACD/UV-IR Manager and UV-IR Processor. Version 6.0 for Microsoft Windows to separate the individual components of the spectrum in the areas corresponding to strongly overlapping absorption bands. The contours shape during the spectra simulation is mixed Gaussian-Lorentzian. After converting the spectra to eliminate the influence of the penetration depth of radiation and automatically determine the main peaks position a preliminary decomposition of the spectrum on these bands was conducted. The first derivative of the experimental contour and the deviation of the calculated spectrum from the experimental spectrum were analyzed. The most probable position of the characteristic peaks not recorded in the preliminary decomposition was determined by the deviation of the resulting deviation from the zero level and the position of the peaks on the graph of the derivative, taking into account the literature data. The addition of a decomposition component was carried out step by step with the expansion of the spectrum conduction, taking into account the added peak and the deviations of the calculated spectrum analysis.

The Melt Flow Rate (MFR) was determined by capillary viscometer IIRT-5M according to GOST 11645-73 under the temperature of 125°C and the load of 2.16 kg.

The intrinsic viscosity was determined by the standard method [5] in carbon tetrachloride solution at 25°C.

9.3 DISCUSSION AND RESULTS

The change of the macromolecules chemical structure as a result of polymer-analogous transformations can be analyzed by various spectroscopic methods. Qualitative differences between the IR spectra of modified and unmodified samples confirm this fact (Figure 1 and 2). On spectra of polymers modified with ethyl silicates the bands related to silicone fragments in the areas of 1020-1090 cm^{-1} and 780-830 cm^{-1}, specific for the stretching vibrations of Si-O, Si-O-Si, and Si-O-C appears, as well as the band of 971 cm^{-1} appears, characterizing Si-O and Si-O-Si bonds in the cross-linked siloxane fragments. The difference in the nature of optical density changes is possibly due to the different structures formed during the reaction of organosilicon fragments.

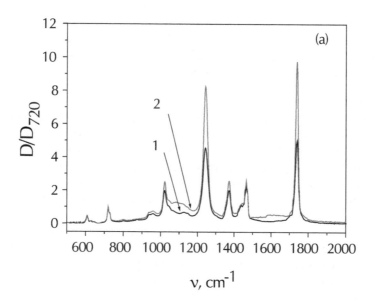

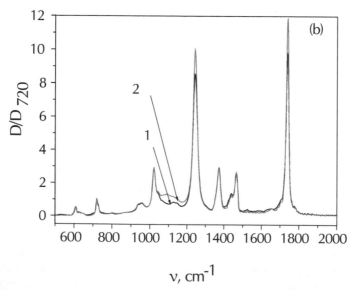

FIGURE 1   IR spectra absorption of EVA 27 (a) and EVAMA 26 (b) initial (1) and modified by ethyl silicates (2).

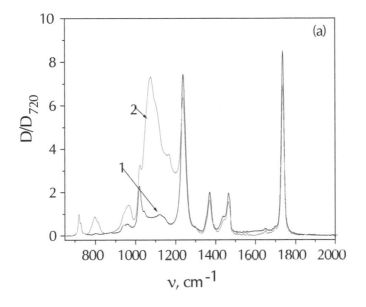

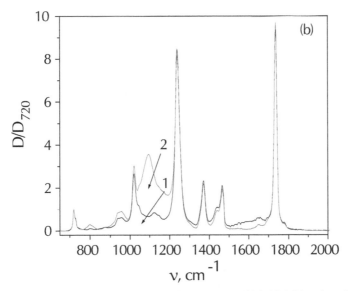

FIGURE 2   IR spectra ATR of EVA 27 (a) and EVAMA 26 (b) initial (1) and modified by cthyl silicates (2).

It was established earlier [1, 2] that the introduction of tetraethoxysilane in the sevilen leads to a splitting in the IR spectra of characteristic band of 1,240 cm^{-1} corresponding to the stretching vibrations of C-O bond in ester groups caused by the substitution of acetyl fragment of the copolymer for the remainder of organosilicon modifier. A similar splitting was observed for ethylene copolymers with vinyl acetate and maleic anhydride (Figure 3). The vibrations of C-O bonds are in the strong interaction with the vibrations of C-C bonds due to the small differences in power coefficients and the proximity of the atoms masses forming the bond. Therefore, the contour of the C-O bands is characterized by the presence of satellites due to rotational isomerism with respect to σ-bonds [6, 7]. The characteristic band shift of the stretching vibrations of C-O bond in the direction of higher frequencies during the substitution of the acetate fragment by the silicone is due to the tension connection by the steric interactions of volume replacement groups [8]. A decrease in the intensity of the characteristic band of 1,462 cm⁻¹ corresponding to the deformation vibrations of methyl groups in vinyl acetate has been observed for EVAMA as well as in the modification of EVA (Figure 4), which confirms the participation of both copolymers EVA and EVAMA in the transesterification reaction of acetyl fragment.

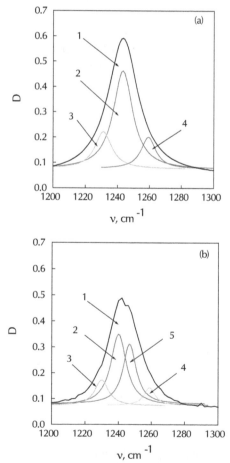

FIGURE 3. Computer decomposition of the infrared spectra in the region of stretching vibrations of C-O bond in ester groups of EVAMA initial (a) and modified by ethyl silicates (b) Observed absorption band of 1,240 cm–1 (1); individual components of the spectrum: 1,240 cm–1 (2), 1230 cm-1 (3), 1,259 cm–1 (4), 1,246 cm–1 (5).

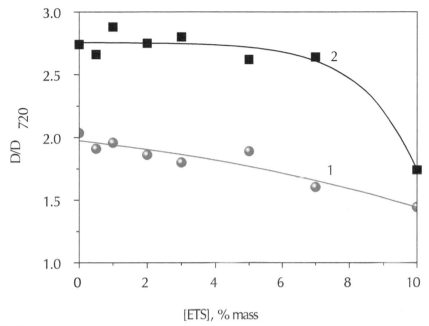

FIGURE 4 Dependence of the relative optical density of the band of 1,462 cm−1 in EVA 27 (1) and EVAMA 26 (2) from the content of ETS (IR-spectroscopy absorption).

The absence of characteristic bands doublets of 1,790 cm^{-1} and 1,850 cm^{-1} related to the stretching vibrations of C=O groups of maleic anhydride, both in the modified and unmodified terpolymer is of great interest, though in IR spectra of films obtained from EVAMA granules, it is presented. We can assume that it is due to the intense thermomechanical effects in the oxygen environment during the rolling, which results in the anhydride cycle opening with the formation of carboxylic acid, reacting further with alkoxy groups of the modifier.

In the spectra of organosilicon compounds the band of 1,070 cm^{-1} corresponding to the stretching vibrations of Si-O-Si groups is indicated vividly. This band presents, respectively, in the spectra of the modified copolymers. The introduction of the modifier into the polymer, naturally, leads to an increase of the optical density of this band corresponding to the total content of Si-O-Si links, which is in the direct ratio to the amount of the injected additive (Figure 5). However, it was found that the surface layers of the polymer are enriched by siloxane phase, as evidenced by the differences in the growth of the characteristic band of 1,070 cm^{-1} in the spectra obtained by ATR (surface layer) and by the absorption IR spectroscopy (integral). It was established earlier [9] that EVA modified by ethylsilicate is a two-component heterophase system. Since, polysiloxanes and polyolefins have different segmental mobility and large differences in free surface energy, it was suggested that the migration of grafted siloxane fragments into the subsurface layers of material is possible during the formation of the samples. The saturation of the polymer surface layer during the silanol modification may be of great interest in the development of adhesive materials.

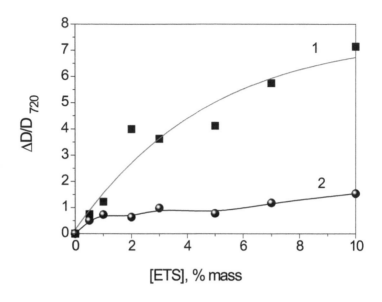

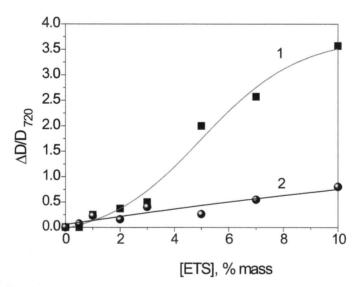

FIGURE 5   Dependence of the relative optical density of the band of 1,070 cm-1 in the ATR spectra (1) and IR spectra absorption (2) in EVA 27 (a) and EVAMA 26 (b) from the content of ETS.

9.4   CONCLUSION

In the course of studies, the effect of maleic anhydride contained in EVAMA on the chemical interaction between polymer and ethyl silicate and the change of the properties of copolymers in the process of silanol modification have not been detected, probably due to the small (not more than 2%) amounts of the comonomer. Previously, it had been found that the introduction of the ethylsilicate into the polymer composites leads to a change in its phase structure [9] and to an increase of the adhesive strength of the material to the metal [1]. Therefore, the data of the effect of ETS on the chemical structure of modified polymers, as well as the fact of the enrichment of the polymers surface layers by the siloxane phase, may be useful in designing of the chemical structure of adhesive materials for different purposes.

KEYWORDS

- **Chemical interaction**
- **Ethylene copolymers**
- **Intrinsic viscosity**
- **IR-spectroscopy**
- **Organosilicon compounds**

REFERENCES

1. *Monomers, Oligomers, Polymers, Composites and Nanocomposites Research. Synthesis, Properties and Applications.* By R. A. Pethrick, P. Petkov, G. E. Zaikov, S. K. Rakovsky (Eds.), Polymer Yearbook, Vol. 23, p. 481 (2012).
2. Stoyanov, O. V., Rusanova, S. N., Petukhova, O. G., and Remisov, A. B. *Russian Applied Chemistry journal,* **74**(7), 1174-1177 (2001).
3. Bounor-Legare, V., Ferreira, I., Verbois, A., Cassagnau, P., and Michel, A. *Polymer,* **43**, 6085-6092 (2002).
4. Bogatyrev, V. L, Maksakova, G. A., Villevald, G. V., and Logvinenko, V. A. Herald of Sibirian Branch of Academy of Sciences of USSR, *Chem. Part,Novosibirsk, Nauka Publisher,* **2**(1), 104-105 (1986).
5. Kurenkov, V. F., Budarina, L. A., and Zaikin, A. E. Physics and Chemistry of Polymers, Moscow, Chemistry-*Koloss Publishing house,* (2008), 256 pp.
6. Barnsa, A. and Tomas, U. Orvill (Eds.) *Vibration spectroscopy: Modern view and tendency.* Mir Publishing house, Moscow, p. 356 (1981).
7. Tomas, Dzh. Orvill (Ed.) *Rotation of molecules.* Mir Publishing house, Moscow, p. 402 (1987).
8. Belopolskaya, T. V. *Russian Polymer Science journal.* A, **14**, 640-645 (1972).
9. Chalykh, A. E., Gerasimov, V. K., Rusanova, S. N., Stoyanov, O. V., Petukhova, O. G., Kulagina, G. S., and Pisarev, S. A. *Russian Polymer Science journal. A.,* **48**(10), 1801-1810 (2006).

10 Updates on Polymer Composites and Fibers for Advanced Technologies

A. K. Haghi and G. E. Zaikov

CONTENTS

10.1 INTRODUCTION

The use of nanostructure materials is not a recently discovered era. Back in the 4th century AD the Romans used nanosized metal particles to decorate cups. The first known and the most famous example is the Lycurgus cup. In this cup, as well as the famous stained glasses of the 10th, 11th, and 12th centuries, metal nanoparticles account for the visual appearance. This property is used in other ways in addition to stained glass. For example, particles of titanium dioxide (TiO_2) have been used for a long time as the sun-blocking agent in sunscreens. Today, nanotechnology refers to technological study and application involving nanoparticles.

In general, nanotechnology can be understood as a technology of design, fabrication, and applications of nanostructures and nanomaterials. Nanotechnology also includes fundamental understanding of physical properties and phenomena of nanometerials and nanostructures. Nanostructure is the study of objects having at least one dimension within the nanoscale. A nanoparticle can be considered as a zero dimensional nanoelement, which is the simplest form of nanostructure.

Many materials are selected for a given application based principally on the material's properties. Most engineering structures are required to bear loads, so the material property of greatest interest is very often its strength. The strength alone is not always enough, however, as in aircraft or many other structures a great penalty accompanies weight. It is obvious an aircraft must be as light as possible, since it must be able to fly.

Although, the terms nanomaterial and nanocomposite represent new and exciting fields in materials science, such materials have actually been used for centuries and have always existed in nature. A nanocomposite is defined as a composite material where at least one of the dimensions of one of its constituents is on the nanometer size scale. The term usually also implies the combination of two (or more) distinct materials, such as a ceramic and a polymer, rather than spontaneously phase segregated structures. The constituent materials will remain separate and distinct at macroscopic level within the finished structure. Generally, two categories of constituent materials, matrix, and reinforcement, exist in the nanocomposite. The matrix materials maintain the relative positions of the reinforcement materials by surrounding and supporting them, and conversely the reinforcements impart their special mechanical or physical properties to enhance the matrix properties. Thus, the composite will have the properties of both matrix and reinforcement, but the properties of a composite are distinct from those of the constituent materials. In recent years, considerable efforts have been devoted for search new functional nanocomposite materials with unique properties that are lacking in their traditional analogues. The control of these properties is an important fundamental problem. The use of nanocrystals as one of the elements of a polymer composite opens up new possibilities for targeted modification of its optical properties because of a strong dependence of the electronic structure of nanocrystals on their sizes and geometric shapes. An increase in the number of nanocrystals in the bulk of composites is expected to enhance long-range correlation effects on their properties. Among the known nanocrystals, nanocrystalline silicon (nc-Si) attracts high attention due to its extraordinary optoelectronic properties and manifestation of quantum size effects. Therefore, it is widely used for designing new generation functional materials for nanoelectronics and information technologies. The use of nc-Si in polymer composites calls for a knowledge of the processes of its interaction with polymeric media.

Solid nanoparticles can be combined into aggregates (clusters), and when the percolation threshold is achieved, a continuous cluster is formed.

An orderly arrangement of interacting nanocrystals in a long-range potential minimum leads to formation of periodic structures. Because of the well developed interface, an important role in such systems belongs to adsorption processes, which are determined by the structure of the nanocrystal surface. In a polymer medium, nanocrystals are surrounded by an adsorption layer consisting of polymer, which may change the electronic properties of the nanocrystals. The structure of the adsorption layer has an effect on the processes of self-organization of solid phase particles, as well as on the size, shape, and optical properties of resulting aggregates. According to data obtained for metallic [1] and semiconducting [2] clusters, aggregation, and adsorption in three-phase systems with nanocrystals have an effect on the optical properties of the whole system. In this context, it is important to reveal the structural features of systems containing nanocrystals, characterizing aggregation, and adsorption processes in these systems, which will make it possible to establish a correlation between the structural and the optical properties of functional nanocomposite systems.

Silicon nanoclusters embedded in various transparent media are a new, interesting object for physicochemical investigation. For example, for particles smaller than 4 nm in size, quantum size effects become significant. It makes possible to control the luminescence and absorption characteristics of materials based on such particles using of these effects [3, 4]. For nanoparticles about 10 nm in size or larger (containing ~10^4 Si atoms), the absorption characteristics in the UV and visible ranges are determined in many respects by properties typical of massive crystalline or amorphous silicon samples. These characteristics depend on a number of factors: the presence of structural defects and impurities, the phase state, and so on. [5, 6]. For effective practical application and creation on a basis nc-Si the new polymeric materials possessing useful properties-sun-protection films [7] and the coverings [8] photoluminescent, and electroluminescent composites [9, 10], stable to light dye [11], embedding of these nanosized particles in polymeric matrixes becomes an important synthetic problem.

The method of manufacture of silicon nanoparticles in the form of a powder by plasma chemical deposition, which was used in this study, makes possible to vary the chemical composition of their surface layers. As a result, another possibility of controlling their spectral characteristics arises, which is absent in conventional methods of manufacture of nc-Si in solid matrices (for example, in SiO_2) by implantation of charged silicon particles [5] or radio frequency deposition of silicon [2]. Polymer composites based on silicon nanopowder are a new object for comprehensive spectral investigation. At the same time, detailed spectral analysis has been performed for silicon nanopowder prepared by laser induced decomposition of gaseous SiH_4 (see, e. g., [6, 12]). It is of interest to consider the possibility of designing new effective UV protectors based on polymer containing silicon nanoparticles [13]. An advantage of this nanocomposite in comparison with other known UV protectors is its environmental safety that is ability to hinder the formation of biologically harmful compounds during UV induced degradation of components of commercial materials. In addition, changing the size distribution of nanoparticles and their concentration in a polymer and correspondingly modifying the state of their surface, one can deliberately change the spectral characteristics of nanocomposite as a whole. In this case, it is necessary to minimize the transmission in the wavelength range below 400 nm (which determines the properties of UV protectors [13]) by changing the characteristics of the silicon powder.

In the first part of this chapter, the possibilities of using polymers containing silicon nanoparticles as effective UV protectors are considered. First, the structure of nc-Si obtained under different conditions and its aggregates, their adsorption and optical properties was studied in order to find ways of control the UV spectral characteristics of multiphase polymer composites containing nc-Si. Also, the purpose of this work was to investigate the effect of the concentration of silicon nanoparticles embedded in polymer matrix and the methods of preparation of these nanoparticles on the spectral characteristics of such nanocomposites. On the basis of the data obtained, recommendations for designing UV protectors based on these nanocomposites were formulated.

The nc-Si consists of core shell nanoparticles in which the core is crystalline silicon coated with a shell formed in the course of passivation of nc-Si with oxygen and/or nitrogen. The nc-Si samples were synthesized by an original procedure in an argon plasma in a closed gas loop. To-do this, we used a plasma vaporizer/condenser operating in a low frequency arc discharge. A special consideration was given to the formation of a nanocrystalline core of specified size. The initial reagent was a silicon powder, which was fed into a reactor with a gas flow from a dosing pump. In the reactor, the powder vaporized at $7,000–10,000°C$. At the outlet of the high temperature plasma zone, the resulting gas-vapor mixture was sharply cooled by gas jets, which resulted in condensation of silicon vapor to form an aerosol. The synthesis of nc-Si in a low frequency arc discharge was described in detail [3].

The microstructure of nc-Si was studied by transmissionelectron microscopy (TEM) on a Philips NED microscope. X-ray powder diffraction analysis was carried out on a Shimadzu Lab XRD-6000 diffractometer. The degree of crystallinity of nc-Si was calculated from the integrated intensity of the most characteristic peak at $2\theta = 28°$. Low temperature adsorption isotherms at 77.3K were measured with a Gravimat-4303 automated vacuum adsorption apparatus. The FTIR spectra were recorded on in the region of $400–5,000$ cm^{-1} with resolution of about 1 cm^{-1}.

Three samples of nc-Si powders with specific surfaces of 55, 60, and 110 m^2/g were studied. The D values for these samples calculated by Equation (2) are 1.71, 1.85, and 1.95, respectively; that is, they are lower than the limiting values for rough objects. The corresponding D values calculated by Equation (3) are 2.57, 2.62, and 2.65, respectively. Hence, the adsorption of nitrogen on nc-Si at 77.3K is determined by capillary forces acting at the liquid-gas interface. Thus, in argon plasma with addition of oxygen or nitrogen, ultra disperse silicon particles are formed, which consist of a crystalline core coated with a silicon oxide or oxynitride shell. This shell prevents the degradation or uncontrollable transformation of the electronic properties of nc-Si upon its integration into polymer media. Solid structural elements (threads or nanowires) are structurally similar, which stimulates self-organization leading to fractal clusters. The surface fractal dimension of the clusters determined from the nitrogen adsorption isotherm at 77.3K is a structurally sensitive parameter, which characterizes both the structure of clusters and the morphology of particles and aggregates of nc-Si.

As the origin materials for preparation film nanocomposites served polyethylene of low density (LDPE) marks 10803-020 and ultradisperse crystal silicon. Silicon powders have been received by a method plazmochemical recondensation of coarse-crystalline silicon in nanocrystalline powder. The synthesis nc-Si was carried out in argon plasma in the closed gas cycle in the plasma evaporator the condenser working in the arc low frequency category. After particle synthesis nc-Si were exposed microcapsulating at which on their surfaces the protective cover from SiO_2, protecting a powder from atmospheric influence and doing it steady was created at storage. In the given work powders of silicon from two parties were used: nc-Si-36 with a specific surface of particles ~36 m^2/g and nc-Si-97 with a specific surface ~97 m^2/g.

Preliminary mixture of polyethylene with a powder nc-Si firms "Brabender" (Germany) carried out by means of closed hummer chambers at temperature 135 ± 5°C, within 10 min and speed of rotation of a rotor of 100 min^{-1}. Two compositions LDPE + nc-Si have been prepared: (1) composition PE + 0.5% nc-Si-97 on a basis nc-Si-97, containing 0.5 weights silicon %; (2) composition PE + 1% nc-Si-36 on a basis nc-Si-36, containing 1.0 weights silicon %.

Formation of films by thickness 85 ± 5 micron was spent on semiindustrial extrusion unit ARP-20-150 (Russia) for producing the sleeve film. The temperature was 120–190°C on zones extruder and extrusion die. The speed of auger was 120 min^{-1}. Technological parameters of the nanocomposites choose, proceeding from conditions of thermostability and the characteristic viscosity recommended for processing polymer melting.

10.2 EXPERIMENTAL

The mechanical properties and an optical transparency of polymer films, their phase structure and crystallinity, and also communication of mechanical and optical properties with a microstructure of polyethylene and granulometric structure of modifying powders nc-Si were observed.

The physico-mechanical properties of films at a stretching (extrusion) measured in a direction by means of universal tensile machine EZ-40 (Germany) in accordance with Russian State Standard GOST-14236-71. Tests are spent on rectangular samples in width of 10 mm, and a working site of 50 mm. The speed of movement of a clip was 240 mm/min. The five parallel samples were tested.

Optical transparency of films was estimated on absorption spectra. Spectra of absorption of the obtained films were measured on spectrophotometer SF-104 (Russia) in a range of wavelengths 200–800 nm. Samples of films of polyethylene and composite films PE + 0.5% nc-Si-36 and PE + 1% nc-Si-36 in the size 3 x 3 cm were investigated. The special holder was used for maintenance uniform a film tension.

The X-ray diffraction analysis by wide angle scattering of monochromatic x-rays data was applied for research phase structure of materials, degree of crystallinity of a polymeric matrix, the size of single-crystal blocks in powders nc-Si and in a polymeric matrix, and also functions of density of distribution of the size crystalline particles in initial powders nc-Si

The X-ray diffraction measurements were observed on Guinier diffractometer: chamber G670 Huber [14] with bent Ge (111) monochromator of a primary beam which are cutting out line $K\alpha_1$ (length of wave $\lambda = 1.5405981$Å) characteristic radiation of X-ray tube with the copper anode. The diffraction picture in a range of corners 2θ from 3 to $100°$ was registered by the plate with optical memory (IP-detector) of the camera bent on a circle. Measurements were spent on original powders nc-Si-36 and nc-Si-97, on the pure film LDPE further marked as PE, and on composite films PE + 0.5% nc-Si-97 and PE + 1.0% nc-Si-36. For elimination of tool distortions effect diffractogram standard SRM660a NIST from the crystal powder LaB_6 certificated for these purposes by Institute of standards of the USA was measured. Further it was used as diffractometer tool function.

Samples of initial powders nc-Si-36 and nc-Si-97 for X-ray diffraction measurements were prepared by drawing of a thin layer of a powder on a substrate from a special film in the thickness 6 μ (MYLAR, Chemplex Industries Inc., Cat. No: 250, Lot No: 011671). Film samples LDPE and its composites were established in the diffractometer holder without any substrate but for minimization of structure effect two layers of a film focused by directions extrusion perpendicular each other were used.

The phase analysis and granulometric analysis was spent by interpretation of the X-ray diffraction data. For these purposes the two different full-crest analysis methods [15, 16] were applied: (1) method of approximation of a profile diffractogram using analytical functions, polynoms and splines with diffractogram decomposition on making parts; (2) method of diffractogram modeling on the basis of physical principles of scattering of X-rays. The package of computer programs WinXPOW was applied to approximation and profile decomposition diffractogram ver. 2.02 (Stoe, Germany) [17], and diffractogram modeling at the analysis of distribution of particles in the sizes was spent by means of program PM2K [18].

10.3 DISCUSSION AND RESULTS

The results of mechanical tests of the prepared materials are presented to Table 1 from which it is visible that additives of particles nc-Si have improved mechanical characteristics of polyethylene.

TABLE 1   Mechanical characteristics of nanocomposite films based of LDPE and nc-Si.

Sample	Tensile strength, kg/cm²	Relative elongation-at-break, %
PE	100 ± 12	$200 - 450$
PE + 1% nc-Si-36	122 ± 12	$250 - 390$
PE + 0.5% nc-Si-97	118 ± 12	$380 - 500$

The results presented in the table show that additives of powders of silicon raise mechanical characteristics of films, and the effect of improvement of mechanical properties is more expressed in case of composite PE + 0.5% nc-Si-97 at which in comparison with pure polyethylene relative elongation-at-break has essentially grown.

Transmittance spectra of the investigated films are shown on Figure 1.

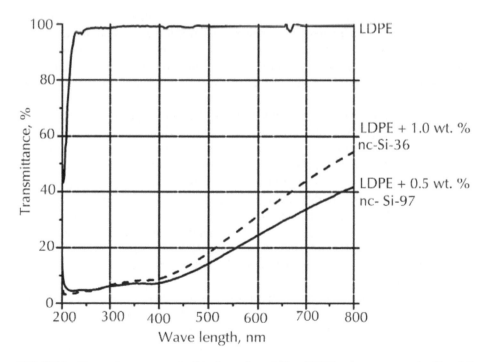

FIGURE 1   Transmittance spectra of the investigated films LDPE and nanocomposite films PE + 0.5% nc-Si-97 and PE + 1.0% nc-Si-36.

It is visible that additives of powders nc-Si reduce a transparency of films in all investigated range of wavelengths but especially strong decrease transmittance (almost in 20 times) is observed in a range of lengths of waves of 220–400 nm, that is in UV areas.

The wide angle scattering of X-rays data were used for the observing phase structure of materials and their component. The measured X-ray diffractograms of initial powders nc-Si-36 and nc-Si-97 on intensity and Bragg peaks position completely corresponded to a phase of pure crystal silicon (a cubic elementary cell of type of diamond–spatial group $Fd\overline{3}m$, cell parameter a_{Si} = 0.5435 nm).

For the present research granulometric structure of initial powders nc-Si is of interest. Density function of particle size in a powder was restored on X-ray diffractogram a powder by means of computer program PM2K [18] in which the method [19] modeling's of a full profile diffractogram based on the theory of physical processes of diffraction of X-rays is realized. The modeling was spent in the assumption of the spherical form of crystalline particles and logarithmically normal distributions of their sizes. The deformation effects from flat and linear defects of a crystal lattice were considered. Received function of density of distribution of the size crystalline particles for initial powders nc-Si are represented graphically on Figure 2, in the signature to which statistical parameters of the found distributions are resulted. These distributions are characterized by such important parameters, as $Mo(d)$ = position of maximum (a distribution mode); $<d>_V$ = average size of crystalline particles based on volume of the sample (the average arithmetic size) and $Me(d)$ = the median of distribution defining the size d, specifying that particles with diameters less than this size make half of volume of a powder.

The results represented on Figure 2, show that initial powders nc-Si in the structure has particles with the sizes less than 10 nm which especially effectively absorb UV radiation. The both powders modes of density function of particle size are very close, but median of density function of particle size of a powder nc-Si-36 it is essential more than at a powder nc-Si-97. It suggests that the number of crystalline particles with diameters is less 10 nm in unit of volume of a powder nc-Si-36 much less, than in unit of volume of a powder nc-Si-97. As a part of a powder nc-Si-36 it is a lot of particles with a diameter more than 100 nm and even there are particles more largely 300 nm whereas the sizes of particles in a powder nc-Si-97 do not exceed 150 nm and the basic part of crystalline particles has diameter less than 100 nm.

The phase structure of the obtained films was estimated on wide angle scattering diffractogram only qualitatively. The complexity of diffraction pictures of scattering and structure do not poses the quantitative phase analysis of polymeric films [20]. At the phase analysis of polymers often it is necessary to be content with the comparative qualitative analysis which allows watching evolution of structure depending on certain parameters of technology of production. Measured wide angle X-rays scattering diffractograms of investigated films are shown on Figure 3. The diffractogramms have a typical form for polymers. As a rule, polymers are the two-phase systems consisting of an amorphous phase and areas with distant order, conditionally named crystals. Their diffractograms represent [20] superposition of intensity of scattering by the amorphous phase which is looking like wide halo on the small angle area (in this case in area 2θ between 10 and 30°), and intensity Bragg peaks scattering by a crystal phase.

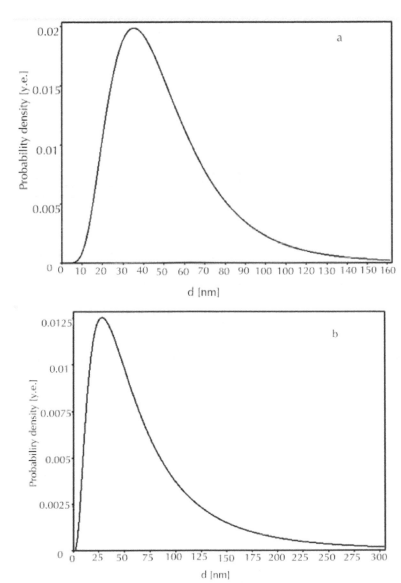

FIGURE 2 Density function of particle size in powders nc-Si, received from x-ray diffractogram by means of program PM2K:

(a) – nc-Si-97	Mo(d) = 35 nm	Me(d) = 45 nm	$<d>_v$ = 51 nm;
(б) – nc-Si-36	Mo(d) = 30 nm	Me(d) = 54 nm	$<d>_v$ = 76 nm.

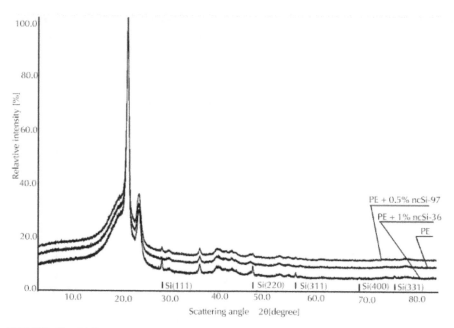

FIGURE 3 Diffractograms of the investigated composite films in comparison with diffractogram of pure polyethylene. Below vertical strokes specify reference positions of diffraction lines of silicon with their interference indexes (hkl).

The data on Figure 3 is presented in a scale of relative intensities (intensity of the highest peak is accepted equal 100%). For convenience of consideration curves are represented with displacement on an axis of ordinates. The scattering plots without displacement represented completely overlapping of diffractogram profiles of composite films with diffractogram of a pure LDPE film, except peaks of crystal silicon which were not present on PE diffractogram. It testifies that additives of powders nc-Si practically have not changed crystal structure of polymer.

The peaks of crystal silicon are well distinguishable on diffractograms of films with silicon (the reference positions with Miller's corresponding indexes are pointed below). The heights of the peaks of silicon with the same name (i.e. peaks with identical indexes) on diffractograms of the composite films PE + 0.5% nc-Si-97 and PE + 1.0% nc-Si-36 differ approximately twice that corresponds to a parity of mass concentration Si set at their manufacturing.

The degree of crystallinity of polymer films (a volume fraction of the crystal ordered areas in a material) in this research was defined by diffractograms Figure 3 for a series of samples only semiquantitative (more/less). The essence of the method of crystallinity definition consists in analytical division of a diffractogram profile on the Bragg peaks from crystal areas and diffusion peak of an amorphous phase [20], as is shown in Figure 4.

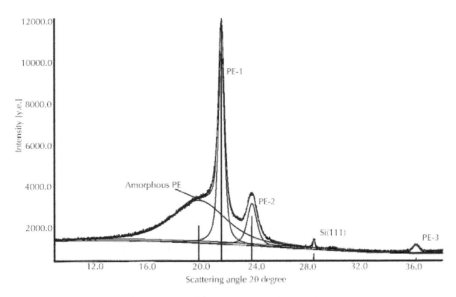

FIGURE 4   Diffractogram decomposition on separate peaks and a background by means of approximation of a full profile by analytical functions on an example of the data from sample PE + 1% nc-Si-36 (Figure 3). The PE-n designate Bragg peaks of crystal polyethylene with serial numbers n from left to right. Si (111)–Bragg silicon peak nc-Si-36. Vertical strokes specify positions of maxima of peaks.

Peaks profiles of including peak of an amorphous phase, were approximated by function pseudo-foigt, a background 4 order polynoms of Chebysheva. The nonlinear method of the least squares minimized a difference between intensity of points experimental and approximating curves. The width and height of approximating functions, positions of their maxima and the integrated areas, and also background parameters were thus specified. The relation of integrated intensity of a scattering profile by an amorphous phase to full integrated intensity of scattering by all phases except for particles of crystal silicon gives a share of amorphy of the sample, and crystallinity degree turns out as a difference between unit and an amorphy fraction.

It was supposed that one technology of film obtaining allowed an identical structure. It proved to be true by coincidence relative intensities of all peaks on diffractograms Figure 3, and samples consist only crystal and amorphous phases of the same chemical compound. Therefore, received values of degree of crystallinity should reflect correctly a tendency of its change at modification polyethylene by powders nc-Si though because of a structure of films they can quantitatively differ considerably from the valid concentration of crystal areas in the given material. The found values of degree of crystallinity are represented in Table 2.

TABLE 2   Characteristics of the ordered (crystal) areas in polyethylene and its composites with nc-Si.

PE			PE + 1% nc-Si-36			PE + 0.5% nc-Si-97		
Crystallinity	**46%**		**47,5%**			**48%**		
2θ [°]	d [E]	ε	2θ [°]	d [E]	e	2θ [°]	d [E]	e
21.274	276	8.9	21.285	229	7.7	21.282	220	7.9
23.566	151	12.8	23.582	128	11.2	23.567	123	11.6
36.038	191	6.8	36.035	165	5.8	36.038	162	5.8
Average values	206	9.5×10^{-3}		174	8.2×10^{-3}		168	8.4×10^{-3}

One more important characteristic of crystallinity of polymer is the size d of the ordered areas in it. For definition of the size of crystalline particles and their maximum deformation ε in X-ray diffraction analysis [21] Bragg peaks width on half of maximum intensity (Bragg lines half-width) is often used. In the given research the sizes of crystalline particles in a polyethylene matrix calculated on three well expressed diffractogram peaks Figure 3. The peaks of polyethylene located at corners 2θ approximately equal 21.28, 23.57, and 36.03° (peaks PE-1, PE-2, and PE-3 on Figure 4 see) were used. The ordered areas size d and the maximum relative deformation ε of their lattice were calculated by the joint decision of the equations of Sherrera and Wilson [21] with use of half-width of the peaks defined as a result of approximation by analytical functions, and taking into account experimentally measured diffractometer tool function. Calculations were spent by means of program $WinX^{POW}$ size/strain. Received d and ε, and also their average values for investigated films are presented in Table 2. The updated positions of maxima of diffraction peaks used at calculations are specified in the Table 2.

The offered technology allowed the obtaining of films LDPE and composite films LDPE + 1% nc-Si-36 and LDPE + 0.5% nc-Si-97 an identical thickness (85 μ). Thus concentration of modifying additives nc-Si in composite films corresponded to the set structure that is confirmed by the X-ray phase analysis.

By direct measurements it is established that additives of powders nc-Si have reduced a polyethylene transparency in all investigated range of lengths of waves, but especially strong transmittance decrease (almost in 20 times) is observed in a range of lengths of waves of 220–400 nm, that is in UV areas. Especially strongly effect of suppression UV of radiation is expressed in LDPE film + 0.5% nc-Si-97 though concentration of an additive of silicon in this material is less. It is possible to explain this fact to that according to experimentally received function of density of distribution of the size the quantity of particles with the sizes is less 10 nm on volume/weight unit in a powder nc-Si-97 more than in a powder nc-Si-36.

Direct measurements define mechanical characteristics of the received films–durability at a stretching and relative lengthening at disrupture (Table 1). The received results show that additives of powders of silicon raise durability of films approximately on 20% in comparison with pure polyethylene. Composite films in comparison with pure polyethylene also have higher lengthening at disrupture, especially this improvement is expressed in case of composite PE + 0.5% nc-Si-97. Observable improvement of mechanical properties correlates with degree of crystallinity of films and the average sizes of crystal blocks in them (Table 2). By results of the x-ray analysis the highest crystallinity at LDPE film + 0.5% nc-Si-97, and at it the smallest size the crystal ordered areas that should promote durability and plasticity increase.

10.4   TEXTILE MATERIALS AND FIBERS

Materials produced from fibers and threads are classified as textile: fabrics, nonwoven materials, fur fabric, carpets, and rugs, and so on. [1]. Textile fibers are the main raw material of the textile industry. According to their origin, these fibers are divided into natural and chemical ones. Natural fibers are plant (cotton, baste fibers), animal (wool, silk) and mineral (asbestos) origin ones.

Chemical fibers are produced from modified natural or synthetic high molecular substances and are classified as artificial ones obtained by chemical processing of natural raw material, commonly cellulose (viscose, acetate), and synthetic ones obtained from synthetic polymers (nylon-6, polyester, acryl, PVC fibers, etc.).

Textile materials are damaged by microorganisms, insects, rodent, and other biodamaging agents. Fibers and fabrics resistance to biodamages primarily depends upon chemical nature of the fibers from which they are made. Most frequently we have to put up with microbiological damages of textile materials based on natural fibers–cotton, linen, and so on, utilized by saprophyte microflora. Chemical fibers and fabrics, especially synthetic ones, are higher biologically resistant, but microorganisms biodegraders can also adapt to them.

Textile material degradation by microorganisms depends on their wear rate, kind and origin, organic composition, temperature, and humidity conditions, degree of aeration, and so on.

With increased humidity and temperature, and restricted air exchange microorganisms damage fibers and fabrics at different stages of their manufacture and application, starting from the primary processing of fibers including spinning, weaving, finishing, and storage, transportation and operation of textile materials and articles from them. Fiber and fabric biodamage intensity sharply increases when contacting with the soil and water, specifically in the regions with warm and humid climate.

Textile materials are damaged by bacteria and microscopic fungi. Bacterial degradation of textile materials is more intensive than the fungal one. The damaging bacterium genus are: *Cytophaga, Micrococcus, Bacterium, Bacillus, Cellulobacillus, Pseudomonas, and Sarcina.* Among fungi damaging textile materials in the air and in the soil the following are detected: *Aspergillus, Penicillium, Alternaria, Cladosporium, Fusarium, Trichoderma,* and so on.

Annual losses due to microbiological damaging of fabrics reach hundreds of millions of dollars [2-10]. Fiber and fabric biodamaging by microorganisms is usually accompanied by the mass loss and mechanical strength of the material as a result, for example, of fiber degradation by microorganism metabolites: enzymes, organic acids, and so on.

10.4.1 Curious Facts

The immediate future of the textile industry belongs to biotechnology. Even today suggestions on the synthesis of various polysaccharides using microbiological methods are present. These methods may be applied to synthesis of fiber-forming monomers and polymers.

Scientists have demonstrated possibilities of microbiological synthesis of some monomers to produce dicarboxylic acids, caprolactam, and so on. Some kinds of fiber-forming polymers, polyethers, in particular, can also be obtained by microbiological synthesis.

Some kinds of fiber-forming polypeptides have already been obtained by the microbiological synthesis. In some cases, concentration of these products may reach 40% of the biomass weight and they can be used as a perspective raw material for synthetic fibers. Studies in this direction are widely performed in many countries all over the world.

The impact of microorganisms on textile materials that causes their degradation is performed, at least, by two main ways (direct and indirect):

• Fungi and bacteria use textile materials as the nutrient source (assimilation);
• Textile materials are damaged by microorganism metabolism (degradation).

Biodamages of textile materials induced by microorganisms and their metabolites manifest in coloring (occurrence of spots on textile materials or their coatings), defects (formation of bubbles on colored surfaces of textile materials), bond breaks in fibrous materials, penetration deep inside (penetration of microorganisms into the cavity of the natural fiber), deterioration of mechanical properties (e.g. strength at break reduction), mass loss, a change of chemical properties (cellulose degradation by microorganisms), liberation of volatile substances and changes of other properties.

It is known that when microorganisms completely consume one part of the substrate they then are able to liberate enzymes degrading other components of the culture medium. It is found that, degrading fiber components each group of microorganisms due to their physiological features decomposes some definite part of the fiber, damages it differently and in a different degree. It is found that along with the enzymes, textile materials are also degraded by organic acids produced by microorganisms: lactic, gluconic, acetic, succinic, fumaric, malic, citric, oxalic, and so on. It is also found that enzymes and organic acids liberated by microorganisms continue degrading textile materials even after microorganisms die. As noted, the content of cellulose, proteins, pectins, and alcohol-soluble waxes increases with the fiber damage degree in it; pH increases, and concentration of water-soluble substances increases that, probably, is explained by increased accumulation of metabolites and consumption of nutrients by microorganisms for their vital activity.

The typical feature of textile material damaging by microorganisms is occurrence of honey dew, red violet or olive spots with respect to a pigment produced by microorganisms and fabric color. As microorganism's pigment interacts with the fabric dye, spots of different hue and tints not removed by laundering or by hydrogen peroxide oxidation. They may sometimes be removed by hot treatment in blankly solution. Spot occurrence on textile materials is usually accompanied by a strong musty odor.

Textile materials are also furnished by biotechnological methods based on the use of enzymes performing various physicochemical processes. Biotechnologies are mostly full for preparation operations (cloth softening, boil-off and bleaching of cotton fabrics, wool washing) and bleaching effects of jeans and other fabrics, biopolishing, and making articles softer.

Enzymes are used to improve sorption properties of cellulose fibers, to increase specific area and volume of fibers, to remove pectin "companions" of cotton and linen cellulose. Enzymes also hydrolyze ether bonds on the surface of polyether fibers.

Cotton fabric dyeing technologies in the presence of enzymes improving coloristic parameters of prepared textile articles under softer conditions, which reduce pollutants in the sewage. To complete treatment of the fabric surface, that is removal of surface fiber fibrilla, enzymes are applied. The enzymatic processing also decreases and even eliminates the adverse effect of long wool fiber puncturing. The application of enzymes to treatment of dyed tissues is industrially proven. This processing causes irregular bleaching that adheres the articles fashionable "worn" style.

Temperature and humidity are conditions promoting biodamaging of fibrous materials. The comparison of requirements to biological resistance of textile materials shall be based on the features of every type of fibers, among which mineral fiber are most biologically resistant.

Different microorganism damaging rate for fabrics is due to their different structure. Thinner fabrics with the lower surface density and higher through porosity are subject to the greatest biodamaging, because these properties of the material provide large contact area for microorganisms and allow their easy penetration deep into it. The thread bioresistance increases with the yarning rate.

10.5 COTTON FIBER BIODAMAGING

Cotton fiber is a valuable raw material for the textile industry. Its technological value is due to a complex of properties that should be retained during harvesting, storage, primary, and further processing to provide high quality of products. One of the factors providing retention of the primary fiber properties is its resistance to bacteria and fungi impacts. This is tightly associated with the chemical and physical features of cotton fiber structures [11].

Ripe cotton fiber represents a unit extended plant cell shaped as a flattened tube with a corkscrew waviness. The upper fiber end is cone-shaped and dead-ended. The lower end attached to the seed is a torn open channel. The cotton fiber structure is formed during its maturation, when cellulose is biosynthesized and its macromolecules are regularly disposed.

The basic elements of cotton fiber morphological structure are known to be the cuticle, primary wall, convoluted layer, secondary wall, and tertiary wall with the central channel [11-14]. The fiber surface is covered by a thin layer of wax substances—the primary wall (cuticle). This layer is a protective one and possesses rather high chemical resistance.

There is a primary wall under the cuticle consisting of a cellulose framework and fatty-wax-pectin substances. The upper layer of the primary wall is less densely packed as compared with the inner one, due to fiber surface expansion during its growth. Cellulose fibrils in the primary wall are not regularly oriented.

The primary layer covers the convoluted layer having structure different from the secondary wall layers. It is more resistant to dissolution, as compared with the main cellulose mass. The secondary cotton fabric wall is more homogeneous and contains the greatest amount of cellulose. It consists of densely packed, concurrently oriented cellulose fibrils composed in thin layers. The fibril layers are spiral-shaped twisted around the fiber axis. There are just few micropores in this layer.

The tertiary wall is an area adjacent to the fiber channel. Some authors think that the tertiary wall contains many pores and consists of weakly ordered cellulose fibrils and plenty of protein admixtures, protoplasm, and pectin substances. The fiber channel is filled with protoplasm residues which are proteins, and contains various mineral salts and a complex of microelements. For the mature fiber, channel cross-section is 4–8% of total cross-section.

Cotton fiber is formed during maturation. This includes not only cellulose biosynthesis, but also ordering of the cellulose macromolecules shaped as chains and formed by repetitive units consisting two β-D-glucose residues bound by glucosidic bonds. Among all plant fibers, cotton contains the maximum amount of cellulose (95–96%).

The morphological structural unit of cellulose is a cluster of macromolecules–a fibril 1.0–1.5 µm long and 8–15 nm thick, rather than an individual molecule. Cellulose fibers consist of fibril clusters uniform oriented along the fiber or at some angle to it.

It is known that cellulose consists not only of crystalline areas–micelles where molecule chains are concurrently oriented and bound by intermolecular forces, but also of amorphous areas.

Amorphous areas in the cellulose fiber are responsible for the finest capillaries formation that is a "sub-microscopic" space inside the cellulose structure is formed. The presence of the sub-microscopic system of capillaries in the cellulose fibers is of paramount significance, because it is the channel of chemical reactions by which water-soluble reagents penetrate deep in the cellulose structure. More active hydroxyl groups interacting with various substances are also disposed here [15-18].

It is known that hydrogen bonds are present between hydroxyl groups of cellulose molecules in the crystalline areas. Hydroxyl groups of the amorphous area may occur free of weakly bound and, as a consequence, they are accessible for sorption. These hydroxyl groups represent active sorption centers able to attract water.

Among all plant fibers, cotton has the highest quantity of cellulose (95–96%). Along with cellulose, the fibers contain some fatty, wax, coloring mineral substances (4–5%). Cellulose concomitant substances are disposed between macromolecule clusters and fibrils. Raw cotton contains mineral substances (K, Na, Ca, Mg) that promote mold growth and also contains microelements (Fe, Cu, Zn) stimulating growth of microorganisms. Moreover, it contains sulfates, phosphorus, glucose, glycidols, and nitrogenous substances, which also stimulate growth of microbes. Differences in their concentrations are one of the reasons for different aggressiveness of microorganisms in relation to the cotton fiber.

The presence of cellulose, pectin, nitrogen-containing, and other organic substances in the cotton fiber, as well as its hygroscopicity makes it a good culture medium for abundant microflora.

Cotton is infected by microorganisms during harvesting, transportation and storage. When machine harvested, the raw cotton is clogged by various admixtures. It obtains multiple fractures of leaves and cotton-seed hulls with humidity higher than of the fibers. Such admixtures create a humid macrozone around, where microorganisms intensively propagate [13]. Fiber humidity above 9% is the favorable condition for cotton fiber degradation by microorganisms.

It is found that cotton fiber damage rate directly in hulls may reach 42–59%; hence, the fiber damage rate depends on a number of factors, for example. cultivation conditions, harvesting period, type of selection, and so on.

Cotton fiber maturity is characterized by filling with cellulose. As the fiber becomes more mature, its strength, elasticity and coloring value increase.

Low quality cotton having higher humidity is damaged by microorganisms to a greater extent. The fifths fibers contain microorganisms 3–5 times more than the first quality fibers. When cotton hulls open, the quantity of microorganisms sharply increases in them, because along with dust wind brings fungus spores and bacteria to the fibers.

Cotton is most seriously damaged during storage: in compartments with high humidity up to 24% of cotton is damaged. Cotton storage in compact bales covered by tarpaulin is of high danger, especially after rains. For instance, after one and half months of such storage the fiber is damaged by 50% or higher.

Cotton microflora remains active under conditions of the spinning industry. As a result, the initial damage degree of cotton significantly increases.

10.5.1 Curious Facts

On some textile enterprises of Ivanovskaya Oblast, sickness cases of spinning-preparatory workshops were observed. When processing biologically contaminated cotton, plenty of dust particles with microorganisms present on them are liberated to the air. This can affect the health status of the employees.

Under natural conditions, cotton products are widely used in contact with the soil (fabrics for tents) receiving damage both from the inside and the outside. The main role here is played by cellulose degrading bacteria and fungi.

It was considered over a number of years that the main role in cotton fiber damage is played by cellulose degrading microorganisms [13]. Not denying participation of cellulose degrading bacteria and fungi in damaging of the cotton fiber, it is noted that a group of bacteria with yellow pigmented mucoid colonies representing epiphytic microflora always present on the cotton plant dominate in the process of the fiber degradation. Nonspore-forming epiphytic bacteria inhabiting in the cotton plants penetrate from their seeds into the fiber channel and begin developing there. Using chemical substances of the channel, these microorganisms then permeate into the submicroscopic space of the tertiary wall primarily consuming pectins of the walls and proteins of the channels.

Enzymes and metabolites produced by microorganisms induce hydrolysis of cellulose macromolecules, increasing damage of internal areas of the fiber. Thus, the fiber delivered to processing factories may already be significantly damaged by microorganisms that inevitably affect production of raw yarn, fabric, and so on.

One hundred thirty five strains of fungi of different genus capable of damaging cotton fibers are currently determined [13]. It is found that the population of phytopathogenic fungi is much lower than that of cellulose degraders: *Chaetomium globosum, Aspergillus flavus, Aspergillus niger, Rhizopus nigricans, and Trichothecium roseum.* These species significantly deteriorate the raw cotton condition sharply reduce spinning properties of the fiber, in particular.

It is also found that the following species of fungi are usually present on cotton fibers: *Mucor* (consumes water-soluble substances), *Aspergillus* and *Penicillium* (consumes insoluble compounds), *Chaetomium, Trichoderma*, and so on (degrade cellulose) [13]. These points to the fact that some species of mold fungi induce the real fiber decomposition that shall be distinguished from simple surface growth of microorganisms. For example, *Mucor* fungi incapable of inducing cellulose degradation may actively vegetate on the yarn finish [14]. Along with fungi, bacteria are always present on the raw cotton, most represented by *Bacillus* and *Pseudomonas* genus species.

Figures 5, 6 present surface micrographs of the first and fifth quality grade primary cotton fibers.

Figure 7, a micrograph, shows the cotton fiber surface after the impact of spontaneous microflora during 7 days. It is observed that bacterial cells are accumulated in places of fiber damage, at clearly noticeable cracks. Figure 8 shows the first quality cotton fiber surface after impact of *Aspergillus niger* culture during 14 days. On the fiber surface mycelium is observed. Figures 5 and 6 show photos of cotton fiber surfaces infected by *Bac. subtilis* (14-day exposure)-bacterial cells on first quality fibers are separated by the surface, forming no conglomerates, whereas on the fifth quality fibers conglomerates are observed that indicates their activity (Figures 8–11).

Academician A. A. Imshenetsky has demonstrated that aerobic cellulose bacteria are able to propagate under increased humidity, whereas fungi propagate at lower humidity. Textile products are destroyed by fungi at their humidity about 10%, whereas bacteria destroy them at humidity level of, at least, 20%. As a consequence, the main attention at cotton processing to yarn should be paid to struggle against fungi, and at wet spinning and finishing fabrics and knitwear not only fungi, but mostly bacteria should be struggled against.

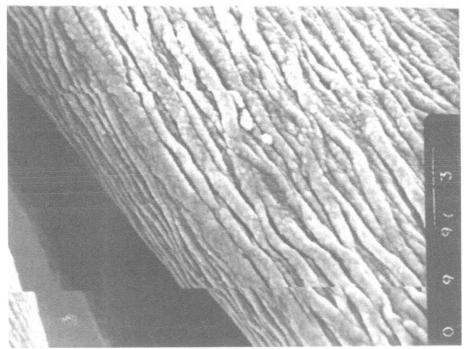

FIGURE 5   The surface of initial first quality cotton specimen (x 4,500).

Cotton damage leads to:

• Significant decrease of strength of the fibers and articles from them;
• Disturbance of technological process (the smallest particles of sticky mucus excreted by some species of bacteria and fungi become the reason for sticking executive parts of machines);
• Abruptness increase;
• Waste volume increase.

Damaging of cotton fibers, fabrics and textile products by microorganisms is primarily accompanied by occurrence of colored yellow, orange, red, violet, and so on spots and then by putrefactive odor and finally the product loses strength and degrades [18-23].

FIGURE 6   The surface of initial fifth quality cotton specimen (x 4,500).

FIGURE 7   The first quality cotton, 7 day exposure (spontaneous microflora) (x 10,000).

FIGURE 8   The first quality cotton, Asp. niger contaminated, 14 day exposure (x 3,000).

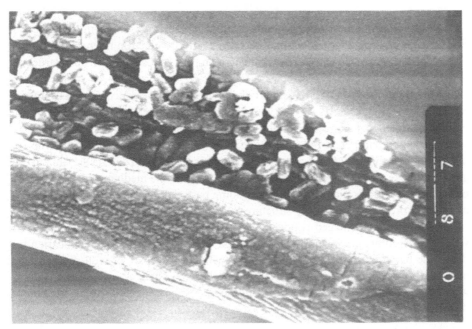

FIGURE 9   The first quality cotton, *Bacillus subtilis* contaminated, 14 day exposure (x 4,500).

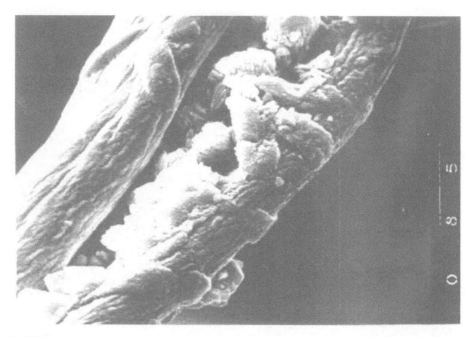

FIGURE 10   The fifth quality cotton, *Bacillus subtilis* contaminated, 14 day exposure (x 4,500).

The effect of microorganisms results in noticeable changes in chemical composition and physical structure of cotton fibers. As found by electron microscopy, cotton fiber degradation by enzymes is most intensive in the zones of lower fibril structure density [24].

In the cotton fiber damaged by microorganisms, cellulose concentration decreases by 7.5%, pectin substances—by 60.7%, hemicellulase—by 20%, and non-cellulose polysaccharides content also decreases. Cellulose biostability increases with its crystallinity degree and macromolecule orientation, as well as with hydroxyl group replacement by other functional groups. Microscopic fungi and bacteria are able to degrade cellulose and as a result glucose is accumulated in the medium, used as a source of nutrition by microorganisms. However, some part of cellulose is not destroyed and completely preserves its primary structure.

Cellulose of undamaged cotton fiber has 76.5% of well ordered area, 7.8% of weakly ordered area, and 15.7% of disordered area. Microbiological degradation reduces the part of disordered area to 12.7%, whereas the part of well ordered area increases to 80.4%. The ratio of weakly ordered area changes insignificantly. This goes to prove that the order degree of cotton cellulose increases due to destruction of disordered areas.

A definite type of fiber degradation corresponds to each stage of the cotton fiber damage. The initial degree of damage is manifested in streakiness, when the fiber surface obtains cracks of different length and width due to its wall break.

Swellings are formed resulting abundant accumulation of microorganisms and their metabolites in a definite part of the fiber. They may be accompanied by fiber wall break induced by biomass pressure. In this case, microorganisms and their metabolites splay out that causes blobs formation from the fiber and breaks in the yarn, as well as irregular fineness and strength.

The external microflora induces the wall damage. The highest degradation stage is fiber decomposition and breakdown into separate fibrils. Hence, perfect fiber structure is absent in this case [13, 25].

In all cases of damage, a high amount of fungal mycelium may be present on the fiber surface, which hyphae penetrate through the fiber or wrap about it thus preventing spinning and coloring of textile materials.

1

FIGURE 11   *(Continued)*

2

3

FIGURE 11   *(Continued)*

4

FIGURE 11   Cotton fiber micrographs: (1) initial fiber (×4,500); *(2–4)* fibers damages by different microorganisms: *(2) Aspergillus niger* (×3,000); *(3) Bacillus subtilis* (×4,500); *(4) Pseudomonas fluorescens* (×10,000).

Enzymatic activity of fungi is manifested in strictly defined places of cellulose microfibrils, and the strength loss rate depends on both external climate conditions and contamination conditions. Cotton fabrics inoculated by the microscopic fungus *Aspergillus niger* under laboratory conditions at a temperature +29°C lose 66% of the initial strength 2–3 weeks after contamination [26], whereas inoculation by *Chaetomium globosum* induces 98.7% loss of strength, that is completely destroys the material.

The same fabric exposed to soil at +29°C during 6 days loses 92% of the initial strength.

And cotton fabric exposed to sea water for 65 days loses up to 90% of strength.

10.6   BAST FIBER BIODAMADING

Fibers produced from stalks, leaves or fruit covers of plants are called bast fibers. Hemp stalks give strong, coarse fibers—the hemp used for packing cloth and ropes. Coarse technical fibers: jute, ambary, ramie, and so on are produced from stalks of cognominal plants. Among all bast fibers, linen ones are most widely used.

The linen complex fiber, from which yarn and fabrics are manufactured, represents a batch of agglutinated filaments (plant cells) stretched and arrow-headed. The linen filament represents a plant cell with thick walls, narrow channel and knee-shaped nodes called shifts. Shifts are traces of fractures or bends of the fiber occurred during growth, and especially during mechanical treatment. Fiber ends are arrow-shaped, and the channel is closed. The cross-section represents an irregular polygon with five or six edges and a channel in the center. Coarser fibers have oval cross-section with wider and slightly flattened channel.

Complex fibers consist of filament batches (15–30 pieces in a batch) linked by middle lamellae. Middle lamellae consist of various substances-pectins, lignin, hemocellulase, and so on.

Bast fibers contain a bit lower amount of cellulose (about 70%) than cotton ones. Moreover, they contain such components as lignin (10%), wax and trace amounts of antibiotics, some of which increase biostability of the fiber. The presence of lignins induces coarsening (lignifications) of plant cells that promotes the loss of softness, flexibility, elasticity, and increased friability of fibers.

The main method for fiber separation from the flax is microbiological one in which vital activity of pectin degrading microorganisms degrade pectins linking bast batches to the stalk tissues. After that the fiber can be easily detached by mechanical processing.

Microorganisms affect straw either at its spreading directly at the farm that lasts 20–30 days or at its retting at a flax-processing plant where retting lasts 2–4 days.

In the case of spreading and retting of spread straw by atmospheric fallouts and dew under anaerobic conditions, the main role is played by microscopic fungi. According to data by foreign investigators, the following fungi are the most widespread at straw spreading: *Pullularia* (spires in the stalk bark); *Cladosporium* (forms a velvet taint of olive to dark green color); *Alternaria* (grows through the bark by a flexible colorless chain and unambiguously plays an important role at dew spreading).

The studies indicate that *Cladosporium* fungus is the most active degrader of flax straw pectins. When retting linen at flax processing plants, conditions different from spreading are created for microflora. Here flax is submerged to the liquid with low oxygen content due to its displacement from straws by the liquid and consumption by aerobic bacteria, which propagate on easily accessible nutrients extracted from the straw.

These conditions are favorable for multiplication of anaerobic, pectin degrading clostridia related to the group of soil spore bacteria, which includes just few species. Most of them are thermophiles and, therefore, the process takes 2–4 days in the warmed up water; however, at lower temperature (+15...20°C) it takes 10–15 days.

10.6.1   Curious Facts

In Russia and Check Republic, spreading is the most popular way of processing flax. In Poland, Romania, and Hungary, the flax is processed at flax processing plants by retting, and I Netherlands–by retting and partly by spreading.

The linen fiber obtained by different methods (spread or retted straw) has different spinning properties. The spread straw is now considered to be the best, where the main role in degradation of stalk pectins is played by mold fungi. In production of retted fiber, this role is played by pectin degrading bacteria, some strains of which being able to form an enzyme (cellulase) that degrades cellulose itself. Such impact may be one of the damaging factors in the processes of linen retting. Thus, biostability of the flax depends on the method of fiber production.

The studies show that all kinds of biological treatment increase the quantity of various microbial damages of the fiber. Meanwhile, the spread fiber had lower total number of microscopic damages compared with any other industrial method.

There are other methods for flax production, steaming, for example, that gives steamed fiber. It has been found that steamed flax is the most biostable fiber. Possible reasons for so high biostability are high structure ordering of this fiber and high content of modified lignin in it. Moreover, during retting and spreading the fiber is enriched with microorganisms able to degrade cellulose under favorable conditions, whereas steaming sterilizes the fiber.

When exposed to microorganisms, pectins content in the linen fiber decreases by 38%, whereas cellulose content–by 1.2% only. The quantity of wax and ash content of the fiber exposed to microorganisms do not virtually change.

The ordered area share in the linen cellulose is 83.6%, the weakly ordered area–5.1%, and disordered area–15.7%. During microbiological degradation the share of disordered areas in the linen cellulose decreases to 7.8%, and the share of ordered areas increases to 86.9%. The share of weakly ordered areas varies insignificantly.

Microbiological damages of linen, jute and other bast fibers and fabrics are manifested by separate staining (occurrence of splotches of color or fiber darkening) and putrefactive odor. On damaged bast fibers, microscopic cross fractures and chips, and microholes and scabs in the fiber walls are observed.

The studies of relative biostability of bast fibers demonstrate that Manilla hemp and jute are most stable, whereas linen and cannabis fibers have the lowest stability.

Natural biostability of bast fibers is generally low and in high humidity and temperature conditions, when exposed to microorganisms, physicochemical and strength indices of both fibers and articles from them rapidly deteriorate. Generally, bast fibers are considered to have virtually the same biostability, as cotton fibers do.

Biostability of cellulose fibers is highly affected by further treatment with finishing solutions (sizing and finishing) containing starch, powder, resins and other substances which confer wearing capacity, wrinkle resistance, fire endurance, and so on. to textile materials. Many of these substances represent a good culture medium for microorganisms. Therefore, at the stage of yarn and fabric sizing and finishing, the main attention is paid to strict compliance with sanitary and technological measures which are to prevent fabric infection by microorganisms and further biodamaging.

10.7 BIODAMAGING OF ARTIFICIAL FIBERS

Artificial fibers and fabrics are produced by chemical treatment of natural cellulose obtained from spruce, pine tree and fir. Artificial fibers based on cellulose are viscose, acetate, and so on. These fibers obtained from natural raw material have higher amorphous structure as compared with high molecular natural material and, therefore, have lower stability, higher moisture, and swelling capacity.

By chemical structure and microbiological stability viscose fibers are similar to common cotton fibers. Biostability of these fibers is low: many cellulosolytic microorganisms are capable of degrading them. Under laboratory conditions, some species of mold fungi shortly (within a month) induces complete degradation of viscose fibers, whereas wool fibers under the same conditions preserve up to 50% of initial stability. For viscose fabrics, the loss of stability induced by soil microorganisms during 12–14 days gives 54–76%. These parameters of artificial fibers and fabrics are somewhat higher than for cotton.

Acetate fibers are produced from acetyl cellulose the product of cellulose etherification by acetic anhydride. Their properties significantly differ from those of viscose fibers and more resemble artificial fibers. For instance, they possess lower moisture retaining property, lesser swelling, and loss of strength under wet condition. They are more stable to damaging effect of cellulosolytic enzymes of bacteria and microscopic fungi, because contrary to common cellulose fibers possessing side hydroxyl groups in macromolecules, acetate fiber macromolecules have side acetate groups hindering interaction of macromolecules with enzymes.

Among artificial textile materials of the new generation, textile fibers from bamboo, primarily obtained by Japanese, are highlighted. Bamboo possesses the reference antimicrobial properties due to the presence of "bambocane" substance in the fiber. Bamboo fibers possess extremely porous structure that makes them much more hygroscopic that cotton. Clothes from bamboo fibers struggle against sweat secretion–moisture is immediately absorbed and evaporated by fabric due to presence of pores, and high antimicrobial properties of bamboo prevent perspiration odor.

Various modified viscose fibers, micromodal, and modal, for example, produced from beech were not studies for biostability. Information on biostability of artificial fibers produced from lactic casein, soybean protein, maize, peanut, and corn is absent.

10.8   WOOL FIBER BIODAMAGING

By wool the animal hair is called, widely used in textile and light industry. The structure and chemical composition of the wool fiber significantly differ it from other types of fibers and shows great variety and heterogeneity of properties. Sheep, camel, goat, and rabbit wool is used as the raw material.

After thorough cleaning, the wool fiber can be considered virtually consisting of a single protein–keratin. The wool contains the following elements (in %): carbon = 50; hydrogen = 6–7; nitrogen = 15–21; oxygen = 21–24; sulfur = 2–5, and other elements.

The chemical feature of wool is high content of various amino acids. It is known that wool is a copolymer of, at least, 17 amino acids, whereas the most of synthetic fibers represent copolymers of two monomers.

Different content of amino acids in wool fibers promotes the features of their chemical properties. Of the great importance is the quantity of cystine containing virtually all sulfur, which is extremely important for the wool fiber properties? The higher sulfur content in the wool is, the better its processing properties are, the higher resistance to chemical and other impacts is and the higher physic-mechanical properties are:

- Wool fiber layers, in turn, differ by the sulfur content: it is higher in the cortical layer that in the core.
- Among all textile fibers, wool has the most complex structure. The fine merino wool fiber consists of two layers:
  - External flaky layer or cuticle and internal cortical layer – the cortex.
  - Coarser fibers have the third layer – the core.

The cuticle consists of flattened cells overlapping one another (the flakes) and tightly linked to one another and the cortical layer inside.

Cuticular cells have a membrane, the so called epicuticle, right around. It is found that epicuticle gives about 2% of the fiber mass. Cuticle cells limited by walls quite tightly adjoin one another, but, nevertheless, there is a thin layer of intercellular protein substance between them, which mass is 3–4% of the fiber mass.

The cortical layer, the cortex, is located under the cuticle and forms the main mass of the fiber and, consequently, defines basic physico-mechanical and many other properties of the wool. Cortex is composed of spindle-shaped cells connivent to one another. Protein substance is also located between the cells.

The cortical layer cells are composed of densely located cylindrical, thread-like macrofibrils of about 0.05–0.2 μm in diameter. Macrofibrils of the cortical layer are composed of microfibrils with the average diameter of 7–7.5 nm [26, 27].

Microfibrils, sometimes call the secondary agents, are composed of primary aggregates–protofibrils. Protofibril represents two or three twisted α-spiral chains.

It is suggested [27, 28] that α-spirals are twisted due to periodic repetition of amino acid residues in the chain, hence, side radicals of the same spiral are disposed in the inner space of another α-spiral providing strong interaction, including for the account of hydrophobic bonds, because each seventh residue has a hydrophobic radical.

According to the data by English investigator J. D. Leeder, the wool fiber can be considered as a collection of flaky and cortical cells bound by a cell membrane complex (CMC) which thus forms a uniform continuous phase in the keratine substance of the fiber. This intercellular cement can easily be chemically and microbiologically degraded, that is a δ-layer about 15 nm thick (CMC or intercellular cement) is located between cells filling in all gaps [29].

The studies show that the composition of intercellular material between flaky cells may differ from that of the material between cortical cells. In the cuticle-cuticle, cuticle-cortex, and cortex-cortex complexes the intercellular "cement" has different chemical compositions.

Although the CMC gives only 6% of the wool fiber mass, there are proofs that it causes the main effect on many properties of the fiber and fabric [29, 30]. For instance, a suggestion was made that CMC components may affect such mechanical properties, as wear resistance and torsion fatigue, as well as such chemical properties, as resistance to acids, proteolytic enzymes and chemical finishing agents.

The core layer is present in the fibers of coarser wool with the core cell content up to 15%. Disposition and shape of the core layer cells significantly vary with respect to the fiber type. This layer can be continuous (along the whole fiber) or may be separated in sections. The cell carcass of the core layer is composed of protein similar to microfibril cortex protein.

By its chemical composition, wool is a protein substance. The main substance forming wool is keratine a complex protein containing much sulfur in contrast with other proteins. Keratine is produced during amino acid biosynthesis in the hair bag epidermis in the hide. Keratine structure represents a complex of high molecular chain batches interacting both laterally and transversally [30-32]. Along with keratine, wool contains lower amounts of other substances.

Wool keratine reactivity is defined by its primary, secondary, and tertiary structures that are the structure of the main polypeptide chains, the nature of side radicals and the presence of cross bonds.

Among all amino acids, only cystine forms cross bonds; their presence considerably defines wool insolubility in many reagents. Cystine bond decomposition simplifies wool damaging by sunlight, oxidants, and other agents. Cystine contains almost all sulfur present in the wool fibers. Sulfur is very important for the wool quality, because it improves chemical properties, strength and elasticity of fibers.

Along with general regularities in the structure of high molecular compounds, fibers differ from one another by chemical composition, monomer structure, polymerization degree, orientation, intermolecular bond strength, and type, and so on. that defines different physico-mechanical and chemical properties of the fibers. The main chemical component of wool–keratine, is nutrition for microorganisms. Microorganisms may not directly consume proteins. Therefore, they are only consumed by microbes having proteolytic enzymes–exoproteases that are excreted by cells to the environment.

Wool damage may start already before sheep shearing that is in the fleece, where favorable nutritive (sebaceous matters, wax, and epithelium), temperature, aeration and humidity conditions are formed. Contrary to microorganisms damaging plant fibers, the wool microflora is versatile, generally represented by species typical of the soil and degrading plant residues.

Initiated in the fleece, wool fiber damages are intensified during its storage, processing and transportation under unfavorable conditions.

Specific epiphytic microflora typical of this particular fiber is always present on its surface. Representatives of this microflora excrete proteolyric enzymes (mostly pepsin), which induce hydrolytic keratine decay by polypeptide bonds to separate amino acids.

Wool is degraded in several stages: first, microorganisms destroy the flaky layer and then penetrate into the cortical layer of the fiber, although the cortical layer itself is not destroyed, because intercellular substance located between the cells is the culture medium. As a result, the fiber structure is disturbed: flakes and cells are not bound yet, the fiber cracks and decays.

The mechanism of wool fiber hydrolysis by microorganisms suggested by American scientist E. Race represents a sequence of transformations: proteins – peptones – polypeptides – water + ammonia + carboxylic acids.

The most active bacteria: *Alkaligenes bookeri, Pseudomonas aeroginosa, Proteus vulgaris, Bacillus agri, B. mycoides, B. mesentericus, B. megatherium, B. subtilis, and microscopic fungi: Aspergillus, Alternaria, Cephalothecium, Dematium, Fusarium, Oospora, Penicillium, and Trichoderma*, were extracted from the wool fiber surface [33–39].

However, the dominant role in the wool degradation is played by bacteria. Fungi are less active in degrading wool. Consuming fat and dermal excretion, fungi create conditions for further vital activity of bacteria-degraders. The role of microscopic fungi may also be reduced to splitting the ends of fibers resulting mechanical efforts of growing hyphae. Such splitting allows bacteria to penetrate into the fiber. Fungi weakly use wool as the source of carbon.

In 1960s, the data on the effect of fat and dirt present on the surface of unclean fibers on the wool biodamaging were published. It is found that unclean wool is damaged much faster than clean one. The presence of fats on unclean wool promotes fungal microflora development. The activity of microbiological processes developing on the wool depends on mechanical damages of the fiber and preliminary processing of the wool.

It is found that microorganism penetration may happen through fiber cuts or microcracks in the flaky layer. Cracks may be of different origins–mechanical, chemical, and so on. It is also found that wool subject to intensive mechanical or chemical treatment is easier degraded by microorganisms than untreated one [39].

For instance, high activity of microorganisms during wool bleaching by hydrogen peroxide in the presence of alkaline agents and on wool washed in the alkaline medium was observed. When wool is treated in a weak acid medium, the activity of microorganisms is abruptly suppressed. This also takes place on the wool colored by chrome and metal-containing dyes. The middle activity of microorganisms is observed on the wool colored by acid dyes.

When impacted by microorganisms, structural changes in the wool are observed: flaky layer damages, its complete exfoliation, and lamination of the cortical layer.

Wool fiber damages can be reduced to several generalized types provided by their structural features:

- Channeling and overgrowth—accumulation of bacteria or fungal hyphae and their metabolites on the fiber surface;
- Flaky layer damage, local and spread;
- Cortical layer lamination to spindle-shaped cells;
- Spindle-shaped cell destruction.

Along with the fiber structure damage, some bacteria and fungi decrease its quality by making wool dirty blue or green that may not be removed by water or detergents. Splotches of color also occur on wool, for example, due to the impact of *Pseudomonas aeruginosa bacteria; in this case, color depends on medium pH: green splotches occurred in a weakly alkaline medium, and in weak acid medium they are red. Green splotches may also be caused by development of Dermatophilus congolensis* fungi. Black color of wool is provided by *Pyronellaea glomerata* fungi.

Thus, wool damage reduces its strength, increases waste quantity at combing and imparts undesirable blue, green or dirty color and putrefactive odor.

However, wool is degraded by microorganisms slower than plant fibers.

10.9   CHANGES OF STRUCTURE AND PROPERTIES OF WOOL FIBERS BY MICROORGANISMS

To evaluate bacterial contamination of wool fibers, it is suggested to use an index suggested by A. I. Sapozhnikova, which characterizes discoloration rate of resazurin solution, a weak organic dye and currently hydrogen acceptor. It is also indicator of both presence and activity of reductase enzyme [40].

The method is based on resazurin ability to lose color in the presence of reductase, which is microorganisms' metabolite, due to redox reaction proceeding. This enables judging about quantity of active microorganisms present in the studied objects by solution discoloration degree.

Discoloration of the dye solution was evaluated both visually and spectrophotometrically by optical density value.

Table 3 shows results of visual observations of color transitions and optical density measurements of incubation solutions after posing the reductase test.

As follows from the data obtained, coloration of water extracts smoothly changed from blue-purple for control sterile physiological solution (D = 0.889) to purple for initial wool samples (D_{thin} = 0.821 and D_{coarse} = 0.779), crimson (D_{thin} = 0.657 and D_{coarse} = 0.651) and light crimson at high bacterial contamination (D_{thin} = 0.548 and D_{coarse} = 0.449 and 0.328) depending on bacterial content of the fibers.

TABLE 3   Visual coloration and optical density of incubation solutions with wool fibers at the wavelength λ = 600 nm and different stages of spontaneous microflora development.

Time of microorganism development, days	Thin merino wool		Coarse caracul wool	
	Optical density	Color (visual assessment)	Optical density	Color (visual assessment)
Control, physiological saline	0.889	Blue purple	0.889	Blue purple
0 (init.)	0.821	Purple	0.779	Purple
7	0.712	Purple	0.657	Crimson
14	0.651	Crimson	0.449	Light crimson
28	0.548	Light crimson	0.328	Light crimson

This dependence can be used for evaluation of bacterial contamination degree for wool samples applying color standard scale [39].

It is found that the impact of microorganisms usually reduces fiber strength, especially for coarse caracul wool: after 28 days of impact strength decreased by 57–65%. The average rate of strength reduction is about 2% per day.

It is found that after 28 days of exposure, the highest reduction of breaking load of wool fibers is induced by *B. subtilis* bacteria [39].

Figure 12 clearly shows that 14 days after exposure to microorganisms the surface of coarse wool fiber is almost completely covered by bacterial cells. Meanwhile, note also (Figure 13) that cuticular cells themselves are not damaged, but their bonding is disturbed that grants access to cortical cells for microorganisms. Figures 14–15 shows the wool fiber decay to separate fibrils caused by microorganisms.

Wool fiber biodegradation changes the important quality indices, such as whiteness and yellowness. This process can be characterized as "yellowing" of wool fibers.

(a)

Obr0 Isx Merinos 9/14/00 SEI   30 µm

(b)

Obr4 isx Grubaya sherst 9/14/00 SEI   30 µm

FIGURE 12 Micrographs of initial wool fibers (×1,000): (a) thin merino wool; (b) coarse caracul wool.

FIGURE 13 Micrographs of thin merino wool (a) and coarse caracul wool (b) after 14 days of exposure to *Bac. subtilis* (×1,000).

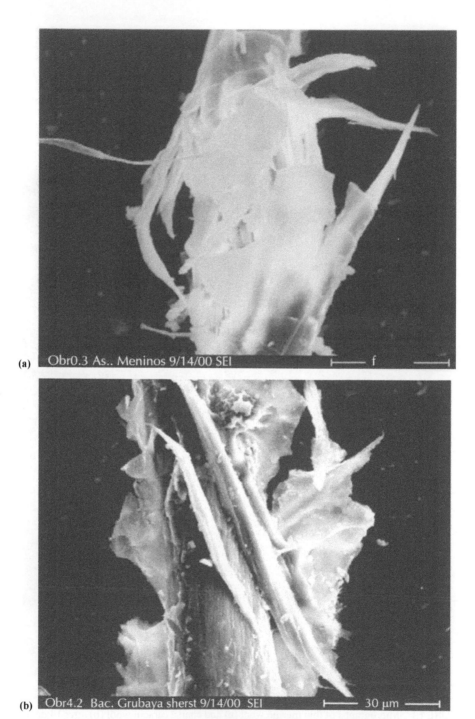

FIGURE 14 Micrographs of flaky layer destruction of thin (a) and coarse and (b) wool fibers after 14 days of exposure to spontaneous microflora (×1000).

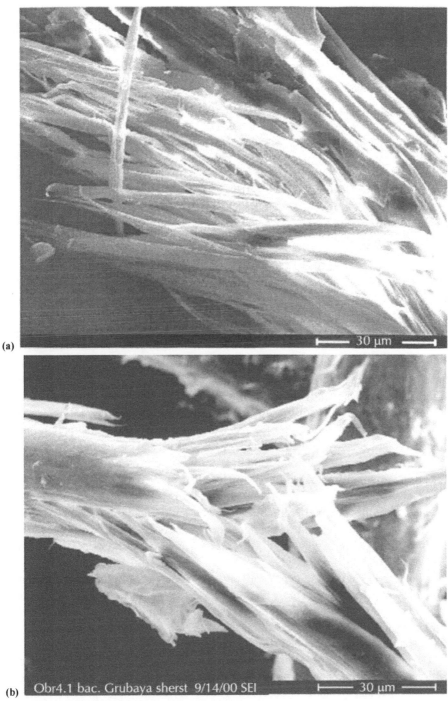

(a)

(b) Obr4.1 bac. Grubaya sherst 9/14/00 SEI

FIGURE 15 Micrographs of thin (a) and coarse and (b) wool fiber fibrillation after 28 days of exposure to microorganisms (×1000).

Table 4 shows data obtained on yellowness of thin merino and coarse caracul wool fibers exposed to spontaneous microflora, *Bac. subtilis* bacteria and *Asp. niger* fungi during 7, 14, and 28 days.

TABLE 4   Yellowness of thin merino and coarse caracul wool fibers after different times of exposure to different microorganisms.

Time of exposure to microorganisms, days		Yellowness, %		
Type of fibers		Impacting microorganisms		
		Bac. subtilis	*Asp. Niger*	Spontaneous microflora
Thin,	0	27.7	27.7	27.7
Thin,	7	36.4	37.3	29.1
Thin,	14	45.4	42.5	34.8
Thin,	28	53.9	48.1	39.9
Coarse	0	39.3	39.3	39.3
Coarse	7	46.8	44.2	41.2
Coarse	14	51.8	46.4	42.1
Coarse	28	57.1	49.5	43.3

In terms of detection of biodegradation mechanism and changes of material properties, determination of relations between changes in properties and structure of fibers affected by microorganisms is of the highest importance. Yellowness increase of wool fibers affected by microorganisms testifies occurrence of additional coloring centers.

For the purpose of elucidating the mechanism of microorganisms' action on the wool fibers that leads to significant changes of properties and structure of the material, the changes of amino acid composition of wool keratine proteins being the nutrition source of microorganisms damaging the material are studied.

Exposure to microorganisms during 28 days leads to wool fiber degradation and, consequently, to noticeable mass reduction of all amino acids in the fiber composition. To the greatest extent, these changes are observed for coarse wool, where total quantity of amino acids is reduced by 10–12 rel. %, and for merino wool slightly lower reduction (4.7 rel. %) is observed.

It should be noted that at comparatively low reduction of the average amount of amino acids in the system (not more than 12 rel. %) all types of wool fibers demonstrate a significant reduction of the quantity of some amino acids, such as serine, cystine, methionine, and son on. (up to 25–33 rel. %).

Analysis of the data obtained testifies that in all types of wool fibers (but to different extent) reduction of quantity of amino acids with disulfide bonds (cystine, methionine) and ones related to polar (hydrophilic) amino acids, including serine, glycine, threonine, and tyrosine, is observed. These very amino acids provide hydrogen bonds imparting stability to keratine structure.

In the primary structure of keratine, serine is the N-end group, and tyrosine is C-end group. In this connection, reduction of the quantity of these amino acids testifies degradation of the primary structure of the protein.

Changes observed in the amino acid composition of wool fiber proteins exposed to spontaneous microflora may testify that microorganisms degrade peptide and disulfide bonds, which provide stability of the primary structure of proteins, and break hydrogen bonds, which play the main role in stabilization of spatial structure of proteins (secondary, tertiary, and quaternary).

Very important data were obtained in the study of the wool fiber structure by IR-spectroscopy method. It is found that when microorganisms affect wool fibers, their surface layers demonstrate increasing quantity of hydroxyl groups that indicates accumulation of functional COO-groups, nitrogen concentration in keratine molecule decreases, and protein chain configuration partly changes β-configuration (stretched chains) transits to α-configuration (a spiral). This transition depends on α- and β-forms ratio in the initial fiber and is more significant for thin fibers, which mostly have β-configuration of chains in the initial fibers.

Thus, it is found that microorganisms generally affect the CMC and degrade amino acids, such as cystine, methionine, serine, glycine, threonine, and tyrosine. Microorganisms reduce breaking load and causes "yellowing" of the wool fibers (Figure 16).

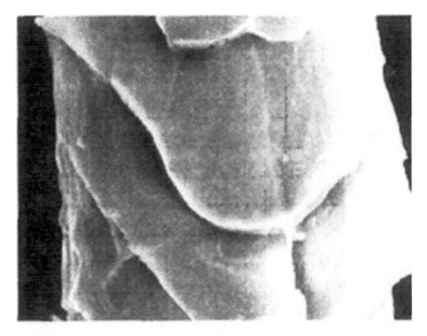

1

FIGURE 16 *(Continued)*

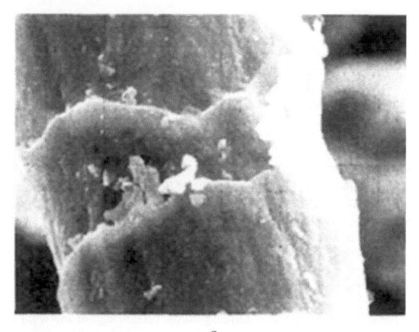

2

3

FIGURE 16   *(Continued)*

4

FIGURE 16   Micrographs of wool fibers: (1) initial undamaged fiber (×3,000); (2, 3) bacterial cells on the fiber surface (×3000); (4) fiber fibrillation after exposure to microorganisms during 4 weeks (×1000).

10.10   BIODAMAGES OF SYNTHETIC FIBERS

Synthetic fibers are principally different from natural and artificial ones by structure and, being an alien substrate for microorganisms, are harder damaged by them. Since occurrence of synthetic fabrics in 1950s, it is suggested that they are "everlasting" and are not utilized by microorganisms. However, it has been found with time that, firstly, microorganisms although slower, but yet are capable of colonizing synthetic fabrics and utilizing their carbon in the course of development (i.e. causing biodamage), and secondly, there are both more and less microorganism resistant fabrics among synthetic ones [41,42].

Among microorganisms damaging synthetic fibers, *Trichoderma* genus fungi are identified, at the initial stages developing due to lubricants and finishing agents without fiber damage and then wrap them with mycelium, loosen threads and, hence, reduce fabric strength.

When studying fabrics from nitrone, lavsan, caprone, it has been found that soil fungi and bacteria cause roughly the same effect on characteristics of these fabrics increasing the fiber swelling degree by 20–25%, reducing strength by 10–15% and elongation at break by 15–20%.

Synthetic fibers represent potential source of energy and nutrition for microorganisms. The ability of microorganisms to attach to surfaces of insoluble solids, then using them as the nutritive substrate, is well-known. Living cells of microorganisms have complex structure, just on the surface of bacterial cells complexes of proteins, lipids and polysaccharides were found; it contains hydrophilic and hydrophobic areas, various functional groups and mosaic electric charge (at total negative charge of the cells).

The first stage of microorganism interaction with synthetic fibers can be rightfully considered in terms of the adhesion theory with provision for the features of structure and properties of microorganisms as a biological system. The entire process of microorganism impact of the fiber can conditionally be divided into several stages: attachment to the fiber, growth and multiplication on it and consumption of it, as the nutrition and energy source [43, 44].

Enzymes excreted by bacteria act just in the vicinity of bacterial membrane. Been adsorbed onto the fiber, living cells attach to the surface and adapt to new living conditions. The ability to be adsorbed onto the surface of synthetic fibers is caused by:

The features of chemical structure of the fibers. For instance, fibers adsorbing microorganisms are polyamide and polyvinyl alcohol ones; the fiber not adsorbing microorganisms is, for example, ftorin;

Physical structure of the fiber. For example, fibers with smaller linear density, with a lubricant on the surface absorb greater amount of microorganisms;

The presence of electric charge on the surface, its value and sign. Positively charged chemical fibers adsorb virtually all bacteria, fibers having no electric charge adsorb the majority of bacteria, and negatively charged fibers do not adsorb bacteria.

Super molecular structure also stipulates the possibility for microorganisms and their metabolites to diffuse inside the internal areas of the fiber. Microorganism assimilation of the fiber starts from the surface, and further degradation processes and their rate are determined by microphysical state of the fiber. Microorganism metabolite penetration into inner areas of the fiber and deep layers of a crystalline material is only possible in the presence of capillaries.

Chemical fiber damages and degradation starting from the surface are, in many instances, promoted by defects like cracks, chips or hollows which may occur in the course of fiber production and finishing.

Along with physical inhomogeneity, chemical inhomogeneity may promote biodegradation of synthetic fibers. Chemical inhomogeneity occurs during polymer synthesis and its thermal treatments, manifesting itself in different content of monomers and various end groups. The possibility for microorganism metabolites to penetrate inside the structure of synthetic fibers depends on the quantity and accessibility of functional end groups in the polymer, which are abundant in oligomers.

The ability to synthetic fibers to swell also makes penetration of biological agents inside low ordered areas of fibers and weakens intermolecular interactions, off orientation of macromolecules, and degradation in the amorphous and crystalline zones. Structural changes result in reduction of strength properties of fibers.

Theoretical statements that synthetic fibers with the lower ordered structure and higher content of oligomers possess lower stability to microorganism impact than fibers with highly organized structure and lower content of low molecular compounds.

Thus, the most rapid occurrence and biodegradation of synthetic fibers are promoted by low ordering and low orientation of macromolecules in the fibers, their low density, low crystallinity and the presence of defects in macro and microstructure of the fibers, pores and cavities in their internal zones.

Carbochain polymer based fibers are higher resistant to microbiological damages. These polymers are-polyolefins, polyvinyl chloride, polyvinyl fluoride, polyacrylonitrile, and polyvinyl alcohol. Fibers based on heterochain polymers: polyamide, polyether, polyurethane, and so on, are less bioresistant.

The comparative soil tests for biostability of artificial and synthetic fibers demonstrate that viscose fiber is completely destroyed on the 17th day of tests; bacterium and fungus colonies occur on lavsan on the 20th day; caprone is overgrown by fungus mycelium on the 30th day. Chlorin and ftorlon have the highest biostability. The initial signs of their biodamage are only observed 3 months after the test initiation.

The studies of nitron, lavsan, and capron fabric biostability have found that soil fungi and bacteria cause nearly equal influence on parameters of these fabrics, increasing swelling degree of the fibers by 20–25%, reducing strength by 10–15% and elongation at break by 15–20%. Meanwhile, nitron demonstrated higher biostability, as compared with lavsan and capron.

10.11   CHANGES IN STRUCTURE AND PROPERTIES OF POLYAMIDE FIBERS INDUCED BY MICROORGANISMS

In contrast with natural fibers, chemical fibers have no permanent and particular microflora. Therefore, the most widespread species of microorganisms possessing increased adaptability are the main biodegraders of these materials.

Occurrence and progression of biological degradation of polyamide fibers is, in many instances, is induced by their properties and properties of affecting microorganisms, and their species composition. Generally, the species of microorganisms degrading polyamide and other chemical fibers are determined by their operation conditions, which form microflora, and its adaptive abilities.

Polyamide fibers are most frequently used in mixtures with natural fibers. Natural fibers contain specific microflora on the surface and inside. Therefore, capron fibers mixed with cotton, wool or linen are affected by their microflora. It is found [43–45] that capron fiber degradation by microorganisms obtained from wool is characterized as deep fiber decay; microorganisms extracted from natural silk cause streakiness of capron fibers; microorganisms extracted from cotton cause fading and decomposition; microorganisms extracted from linen cause fading, streakiness and decomposition.

The microorganism interaction with polyamide fibers is most fully studied in the works by I. A. Ermilova [43-45]. For the purpose of detecting bacteria-degraders of polyamide fibers, microorganisms were extracted from fibers damaged in the medium of active sewage silt, soil, microflora of natural fibers and test-bacteria complex selected as degraders of polyamide materials. Capron fibers were inoculated by extracted cultures of microorganisms and types of damages were reproduced.

Polyamide fiber materials were natural nutritive and energy source for these microorganisms. Therefore, bacterial strains extracted from damaged fibers were different from the initial strains. It is proven that the existence on a new substrate has stipulated changes of intensity and direction of physiological-biochemical processes of bacterial cells, the change of their morphological and culture properties [43-45].

Extraction, cultivation and use of such adaptive strains are of both scientific and practical interest. Using bacterial strains adaptive to capron, polyamide production waste, warn products, toxic substances may be utilized. This allows obtaining of secondary raw materials and solving the problem of environmental protection.

It is proved experimentally [43-45] that polycaproamide (PCA) fibers possess high adsorbability, which value depends on the properties of impacting bacteria. For instance, gram-positive, especially spore-forming bacteria *Bacillus subtilis, B. mesentericus* (from 84.5 to 99.3% of living cells), are most highly adsorbed, and adsorption of gram-positive bacteria varies significantly.

The extensive research of test-bacterium complex impact was performed [43-45]. These bacteria were chosen as degraders of polyamide fibers, along with microflora of active sewage silts, linen, and jute microorganisms as degraders of a complex capron thread. It is found that the test-bacterium complex injected by the author, after 7 months of exposure, increases biodegradation index to 1.27 that testifies about intensive degradation of the fiber microstructure. The highest degradation of complex PCA thread is caused by from active sewage silt microorganisms.

High activity of silt microorganisms and test-bacteria is explained by the fact that they include bacterium strains *Bacillus subtilis* and *B. mesentericus*, which according to the data by a number of authors [39] may induce full degradation of caprolactam to amino acids using it, as the source of carbon and nitrogen.

Thus, adaptive forms of microorganisms induce the highest degradation of polyamide fibers.

Polyamide fibers are characterized by physical structure inhomogeneity, which occurs during processing and is associated with differences in crystallinity and orientation of macromolecules determining fiber accessibility for microorganisms and their metabolites penetration. The surface layer is damaged during orientational stretching and, consequently, has lower molecular alignment. That is why the surface layer is most intensively changed by microorganisms. The study of polyamide fiber macrostructure after exposure to microorganisms shows that streakiness and cover damage are the main damages of these fibers [43].

The studies of supermolecular structure of capron fiber surface show that after exposure to microorganisms the capron fiber cover becomes loose and uneven [39, 43]. Surface super molecular structure degradation increases with the microbial impact: fibrils and their yarns become split and disaligned both laterally and transversely, multiple defects in the form of pores and cavities, and cracks of various depths are formed.

Chemical in homogeneity of polyamide fibers also promotes changes in the fiber structure, when exposed to microorganisms [46-50]. It is found that the polyamide fiber degradation increases with low molecular compound (LMC) content in them; meanwhile, at the same content of low-molecular compounds, thermally treated fibers were higher biostable, as compared with untreated specimens [43, 44].

Along with the morphological characters which characterize biodamaging of the fibers, functional features, such as strength decrease and increase of deformation properties of the fiber, were detected. The greatest strength decrease (by 46.4%) was observed for thermally untreated capron fiber containing 3.4% LMC, and the smallest decrease (by 5%) was observed for the fibers with 3.2% LMC, thermally treated at the optimum time of 5 s [43, 44].

The IR-spectroscopy method was applied to detect PCA fiber damage by microorganisms [39]. It is found that carboxyl and amide groups are accumulated during their biodegradation.

The change of various property indices of polyamide fibers also results from macro, micro, and chemical changes.

Of interest are studies of the microorganism effect on polyamide fabric quality [43, 44]. Fabrics (both bleached and colored) from capron monofilament were exposed to a set of test-cultures: *Bac.subtilis, Ps.fluorecsens, Ps.herbicola,* and *Bac.mesentericus.* After 3–9 month exposure, yellowness and dark spots occurred, coloring intensity decreased, and an odor appeared. The optical microscopy studies indicated that all fibers exposed to microorganisms, had damages typical of synthetic fibers overgrowing, streakiness, bubbles, wall damages. The increasing quantity of biodamages with time results in tensile strength reduction: by 6–8% for capron fibers after 9 months, at inconsiderable change of relative elongation.

It all goes to show that development of microorganisms on polyamide fibrous materials results in changes of fibers morphology, their molecular and super molecular structure and, as a consequence, reduction of strength properties, color change, and odor.

To clear up the mechanism of PCA fiber degradation by polarographic investigation, a possibility of ε-amino caproic acid (ACA) accumulation by *Bacillus subtilis k1* culture during degradation of PCA fibers 0.3 and 0.7 tex fineness was studied (Figure 17) [39].

As a source of carbon, ACA (10 mg/l) or PCA (0.5 g/l) was injected into the mineral medium.

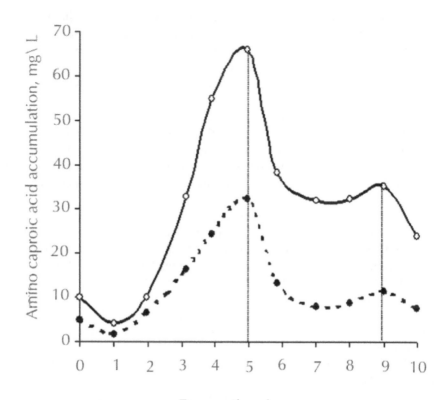

Exposure time, days

FIGURE 17   The change of ACA concentration during Bacillus subtilis k1 development on PCA material fibers: –0.7 tex; –0.3 tex.

It is found that at PCA fibrous materials exposure to *Bacillus subtilis k1* strain, the maximum quantity of ACA is liberated on the fifth day: 32 mg/l, 66 mg/l.

Amino caproic acid accumulation, mg\l

It is known that if the medium includes several substrates metabolized by a particular strain of microorganisms, the substrate providing the maximum culture propagation rate is consumed first. As this substrate is going to exhaust, bacteria subsequently consume other substrates, which provide lower rates of cell multiplication.

In the mineral medium containing chemically pure ACA (with the initial concentration of 10 mg/l) as the source of carbon, its concentration decreased gradually, and 5 days after it was not detected in the solution.

Thus, basing on the data obtained, one may conclude that, firstly, PCA fibers of 0.3 tex fineness are more accessible for microorganisms; secondly, *Bacillus subtilis k1* strain can be used to utilize PCA fibrous materials; thirdly, polarographic analysis has proven the mechanism of PCA fibrous material degradation with ACA liberation. As a consequence, suggested *Bacillus subtilis k1* strain VKM No.V-1676D degrades PCA fibrous materials at the both macro- and microstructure levels, with ACA formation.

10.12   METHODS OF TEXTILE MATERIAL PROTECTION AGAINST DAMAGING BY MICROORGANISMS

Imparting antimicrobial properties to textile materials pursues two main aims: protection of the objects contacting with textile materials against actions of microorganisms and pathogenic microflora.

In the first case, we speak about imparting biostability to materials and, as a consequence, about passive protection. The second case concerns creation of conditions for preventive attack of a textile material on pathogenic bacteria and fungi to prevent their impact on the protected object [51].

The basic method of increasing biostability of textile materials is application of antimicrobial agents (biocides). The requirements to the "ideal" biocide are the following:

The efficacy against the most widespread microorganisms at minimal concentration and maximal action time;

Non-toxicity of applied concentrations for people;

The absence of color and odor;

Low price and ease of application;

Retaining of physicomechanical, hygienic and other properties of the product;

Compatibility with other finishing agents and textile auxiliaries;

Light stability and weatherability.

At any time, nearly every class of chemical compounds was applied to impart textile materials antibacterial or antifungal activity. Today, application of nanotechnologies, specifically injection of silver and iodine nanoparticles, to impart textile antimicrobial properties is of the greatest prospect.

At all times, copper, silver, tin, mercury, and so on. salts were used to protect fibrous materials against biodamages. Among these biocides, the most widespread are copper salts due to their low cost and comparatively toxicity. The use of zinc salts is limited by their low biocide action, whereas mercury, tin and arsenic salts are highly toxic for the man [52-57]. However, there are organomercury preparations applied to synthetic and natural fibrous materials used as linings and shoe plates, widely advertised for antibacterial and antifungal finishing.

The data are cited [52-57] that impregnation of textile materials by a mixture of neomycin with tartaric, propionic, stearic, phthalic, and some other acids imparts them the bacteriostatic effect. Acids were dissolved in water, methyl alcohol or butyl alcohol and were sprayed on the material.

The methods of imparting textile materials biostability can be divided into the following groups:

Impregnation by biocides, chemical and physical modification of fibers and threads, which then form a textile material;

Cloth impregnation by antimicrobial agent solutions of emulsions, its chemical modification;

Injection of antimicrobial agents into the binder (at nonwoven material manufacture by the chemical method);

Imparting antimicrobial properties to textile materials during their coloring and finishing;

Application of disinfectants during chemical cleaning or laundry of textile products.

However, impregnation of fibers and cloth does not provide firm attachment of reagents. As a result, the antimicrobial action of such materials is nondurable. The most effective methods of imparting biocide properties to textile materials are those providing chemical bond formation that is chemical modification methods. Chemical modification methods for fibrous materials represent processing that leads to clathrate formation, for example injection of biological active agents into spinning melts or solutions.

At the stage of capron polymerization, an antibacterial organotin compound (tributyltin oxide or hydroxide) is added that retains the antibacterial effect after multiple laundries. Methods of imparting antimicrobial properties to textile materials by injection of nitrofuran compounds into spinning melts with further fixing them at molding in the fine structure of fibers similar to clathrates were designed.

There are data on imparting antimicrobial properties to synthetic materials during oiling. Prior to drafting, fibers are treated by compounds based on oxyquinoline derivatives, by aromatic amines or nitrofuran derivatives. Such fibers possess durable antimicrobial effect [58-60].

Nanotechnologies are actively intruded in the light industry, allowing obtaining of materials with antimicrobial properties. The following directions using nanotechnologies, which are now investigated, should be outlined, including creation of new textile materials on the account of:

Primarily, the use of textile nonfibers and threads in the materials;

Secondly, the use of nanodispersions and nanoemulsions for textile finishing.

It may be said that nanotechnologies allow a significant decrease of expenses at the main production stage, where consumption of raw materials and semiproducts is considerable. For nanoparticles imparting antimicrobial properties, silver, copper, palladium, and so on. particles are widely used. Silver is the natural antimicrobial agent, which properties are intensified by the nanoscale of particles (the surface area sharply increases), so that such textile is able to kill multiple microorganisms and viruses. In this form, silver also reduces necessity in fabric cleaning, eliminates sweat odor as a result of microorganism development on the human body during wearing.

Properties of materials designed with the use of silver nanoparticles, which prevent multiplication of various microorganisms, may be useful in medicine, for example. The examples are surgical retention sutures, bandages, plasters, surgical boots, medical masks, whites, skull-caps, towels, and so on. Customers know well sports clothes, prophylactic socks with antimicrobial properties.

Along with chemical fibers and threads, natural ones are treated by nanoparticles. For example, silver and palladium nanoparticles (5–20 nm in diameter) were synthesized in citric acid, which prevented their agglutination, and then natural fibers were dipped in the solution with these negatively charged particles. Nanoparticles imparted antibacterial properties and even ability to purify air from pollutants and allergens to clothes and underwear.

When these products appeared at the world market, disputes about ecological properties and the influence of these technologies on the human organism have arisen. There are no accurate data yet how these developments may affect the human organism. However, it should be noted that some specialists do not recommend everyday use of antibacterial socks, because these antibacterial properties affect the natural skin microflora.

Nanomaterials are primarily hazardous due to their microscopic size. Firstly, owing to small size they are chemically more active because of a great total area of the nanosubstance. As a result, low toxic substance may become extremely toxic. Secondly, chemical properties of the nanosubstance may significantly change due to manifestations of quantum effects that, finally, may make a safe substance extremely hazardous. Thirdly, due to small size, nanoparticles freely permeate through cellular membranes damaging bioplasts and disturbing the cell operation.

Physical modification of fibers or threads is the direct change of their composition (without new chemical formations and transformations), structure (supermolecular and textile), properties, production technology, and processing. Modernization of the structure and increase of the fiber crystallinity degree induces biostability increase. However, in contrast with chemical modification, physical modification does not impart antimicrobial properties to the fibers, but may increase biostability.

By no means always textile materials produced completely from antimicrobial fibers are required. Even a small fracture of highly active antimicrobial fiber (e.g. 1/3 or even 1/4) is able to provide sufficient biostability to the entire material. The studies show that antimicrobial fibers were found not only protected themselves against microorganism damage, but also capable of shielding plant fibers from their impact.

Manufacture of antimicrobial nonwoven materials by injection of active microcapsule ingredients into it is of interest. Microcapsules can contain solid particles of microdrops of antimicrobial substances liberated under particular conditions (e.g. by friction, pressure, dissolution of capsule coatings or their biodegradation).

Biostability of fibrous materials may be significantly affected by the dye selection. Dyes possessing antimicrobial activity on the fiber are known: salicylic acid derivatives capable of bonding copper, triphenylmethane, acridic, thiazonic, and so on. ones. For instance, chromium-containing dyes possess antibacterial action, but resistance to mold fungi is not imparted.

It is known that synthetic fibers dyed by dispersed pigments are more intensively degraded by microorganisms. It is suggested that these pigments make the fiber surface more accessible for bacteria and fungi.

Single bath coloring and bioprotective finishing of textile materials are also applied. A combination of these processes is not only of theoretical interest, but is also perspective in terms of technology and economy.

Processing of textile materials by silicones also imparts antimicrobial properties to these clothes. Some authors state that textile material sizing by water repellents imparts them sufficient antimicrobial activity. Water repellency of materials may reduce the adverse impact of microorganisms, because the quantity of adsorbed moisture is reduced. However, Hydrophobic finishing itself may not fully eliminate the adverse effect of microorganisms. Therefore, antimicrobial properties imparted to some textile materials during silicone finishing may be related to application of metal salts as catalysts, such as copper, chromium and aluminum.

Disinfectants, for example at laundry, may be applied by the customer himself. The method of sanitizing substance application for carpets, which is spraying or dispensing of a disinfectant on the surface of floor covers during operation is known. The acceptable disinfection level may be obtained during laundry of textile products by such detergents, which may create residual fungal and bacteriostatic activity [58-60].

KEYWORDS

- **Diffractogram**
- **Fibers and fabrics**
- **Nanocomposite**
- **Nanocrystalline silicon**
- **Nanotechnology**
- **Textile materials**
- **UV protectors**

REFERENCES

1. Pekhtasheva, E. L. *Biodamages and protections of non-food materials*. Masterstvo Gogolia (Ed.). Publishing House, Moscow, p. 224 (in Russian) (2011).
2. Emanuel, N. M. and Buchachenko, A. L. *Chemical physics of degradation and stabilization of polymers*. VSP International Science Publ., Utrecht, p. 354 (1982).
3. Zaikov, G. E., Buchachenko, A. L., and Ivanov, V. B. *Aging of polymers, polymer blends and polymer composites*. Nova Science Publ., New York, Vol. 1, p. 258 (2002).
4. Zaikov, G. E., Buchachenko, A. L., and Ivanov, V. B. *Aging of polymers, polymer blends and polymer composites*, Nova Science Publ., New York, Vol. 2, p. 253 (2002).
5. Zaikov, G. E., Buchachenko, A. L., and Ivanov, V. B. *Polymer aging at the cutting adge*. Nova Science Publ., New York, p. 176 (2002).
6. Gumargalieva, K. Z. and Zaikov, G. E. *Biodegradation and biodeterioration of polymers. Kinetical aspects*. Nova Science Publ., New York, p. 210 (1998).
7. Semenov, S. A., Gumargalieva, K. Z., and Zaikov, G. E. *Biodegradation and durability of materials under the effect of microorganisms*. VSP International Science Publ., Utrecht, p. 199 (2003).
8. Polishchuk, A. Ya. and Zaikov, G. E. *Multicomponent transport in polymer systems*. Gordon and Breach, New York, p. 231 (1996).
9. Moiseev, Yu. V. and Zaikov, G. E. *Chemical resistance of polymers in reactive media*. Plenum Press, New York, p. 586 (1987).

10. Jimenez, A. and Zaikov, G. E. *Polymer analysis and degradation*. Nova Science Publ., New York, p. 287 (2000).

11. Babaev, D. The quality of production. *Khlopkovodstvo (Cotton Production) journal*, (6), pp. 14–18 (Russian) (2009).

12. Ermilova, I. A. and Semenova, D. I. Investigation of bioresistant properties of cotton fibres. *Tekstil'naya Promyshlennost' (Textile Industry) journal*, (4), 13–14 (Russian) (2011).

13. Ipatko, L. I. *Effect of microorganizmes on the structure and properties of cotton fibers*. Thesis, Leningrad Institute of Building Technology, Leningrad, p. 143 (1988).

14. Aminov, Kh. A. Measurement of fibers quality of cotton. *Khlopkovaya Promyshlennost' (Cotton Idustry) journal*, (6), 3, 4 (Russian) (2010)

15. Guban, I. N., Voropaeva, N. L., Rashidova, S. Sh. Paisting of cotton fibers. Doklady Academy of Science Yuzbek. *USSR (Reports of Yuzbek. Academy of Sciences) journal*, (12), 48–50 (Russian) (1988).

16. David, T. W. Chun. High moisture storage effects on cotton stickiness. *Text. Res. J.*, **68**(9), 642–648 (1998).

17. Evans Elaine and Brain Mc Carthy. Biodeterioration of natural fiber. *J. Soc. Dyers Colour.*, **114**(4), 114–116 (1998).

18. Mangialardi, Y. J., Lalor, W. F., Bassett, D., and Miravalle, R. J. Influence of Yrowth Period on neps in Cotton. *Text. Res. J.*, **57**(7) 421–427 (1987).

19. Perkins, H. H. Spin Finishes for Cotton. *Text. Res. J.*, **58**(3) 173–179 (1988).

20. Rakhimov, A. *Investigation of durability and degradation of cotton fibers*. Donish Publishing House, Dushanbe, p. 247 (Russian) (1971).

21. Xu, B., Fang, C., and Watson, M. D. Investigation new factors in cotton color grading. *Text. Res. J.*, **68**(11) 779–787 (1998).

22. Bose, R. G. and Ghose, S. N. Detection of Mildew growth on jute and cotton textiles by ultraviolet light. *Text. Res. J.*, **39**(10) 982–983 (1969).

23. Kaplan, A. M., Mandels, M., and Greenberger, N. Mode of action of regins in preventing microbial degradation of cellulosic textiles. In *Biodeterioration of materials–Volume 2*. pp. 268–278 (1972).

24. Abu-Zeid, A. and Abou-Zeid. A technique for measuring microbial damage of cellulosic sources by microorganisms. *Pakistan J. Sci.*, **23**(½) 21–25 (1971).

25. Piven, T. V. and Khodyrev, V. I. Biodegradation of flax and cotton. *Khimiya Drevesiny (Chemestry of wood) journal*, (1) 100–105 (Russian) (1988).

26. Fucumura, T. Hydrolysis of cyclic and liner oligomers of 6-aminocaproic acid by a bacterial cell extract. *J. of Biochemistry*, **59** 531–536 (1966).

27. Alexander, P. and Hudson, R. F. *Physics and Chemistry of wool*. Chimiya (Chemistry) Publishing House, Moscow, p. 58 (Russian) (2010).

28. Novorodovskaya, T. E. and Sadov, S. F. *Chemistry and chemical technology of wool*. Lesprombytizdat (Forest-industry Publishing House), Moscow, p. 245 (Russian) (1986).

29. Andrzej Wlochowicz and Anna Pielesz. Struktura wlokien welnianych w swetla aktualnych badan. *Prz. Wlok.*, (4) C4–8 (2009).

30. Leeder, J. D. The cell membrane complex and its influence on the properties of the wool fiber. International Wool Secretariat. Development Center. *Wool science review*, **63** 3–35 (1986).

31. Lewis, J. *Microbial biodeterioration. Economic Microbiology*. A. H. Rose (Ed.)., Academic Press, London, pp. 81–130 (1981).

32. Brian J. Mc Carthy. Biodeterioration in wool textile processing. *International Dyer*, **164**, 59–62 (1980).

33. Onions, W. J. *Wool an introduction to its properties, varieties, uses and production*. Interscience, p. 41 (1962).

34. Brian J. Mc Carthy and Phil H. Greavest. Mildew causes, detection methods and prevention. *Wool Sci.Rev.*, **65** 27–48 (1988).

35. Lewis, J. Mildew proofing of wool in relation to modern finishing techniques. *Wool Sci. Rev.*, **1**(46) 17–29(1973).

36. Lewis, J. Mildew proofing of wool in relation to modern finishing techniques. *Wool Sci. Rev.*, 2(47) 17–23 (1973).
37. Jain, P. C. and Agrawal, S. C. A not on the keratin decomposing capability of some fungi. *Transactions of the Mycology Society of Japan*, 21 513–517 (1980).
38. Espie, S. A. and Manderson, G. J. Correlation of microbial spoilage of woolskins with curing treatments. *Journal of Applied Bacteriology*, 47 113–119 (1979).
39. Evans Elaine and Braian Mc. Carthy. Biodeterioration of natural fibers. *J. Soc. Dyers Colour*, 114(4),. 114–116 (1998).
40. Pekhtasheva, E. L., Sapozhnikova, A. I., Neverov, A. N., and Sinitsin N. M. Estimation of amount of microbes in wool fibers. Izvestiya Vuzov. *Tekhnologiya Tekstil'noi Promyshlennosti (Herald of High School. Technology of Textile Industry) journal*, 2(271) (Russian) (2009).
41. Kato, K. and Fukumura, T. Bacterial breakdown of ε–caprolactam. *Chem. and Industr*, 23 1146 (2010).
42. Huang, S. J., Bell, J. P., and Knox, J. R. Desing, Synthesis and Degradation of Polymers Susceptible to Hydrolysis by Proteolytic Enzymes. *Proceeding of Third International Biodegradation Symposium.* London Appl. Sci. Publ. Ltd, Kingston, USA, pp. 731–741 (1975).
43. Ermilova, I. A. *Theoretical and practical foundation of microbiological degradation of textile fibers and ways of defence of fibers against the action of microorganizmes.* Phesis (Russian), S. M. Kirov Leningrad Institute of Textile and Light Industry, Leningrad, p. 470 (2009).
44. Ermilova, I. A. *Theoretical and practical foundation of microbiological degradation of chemical fibers.* Nauka (Science) Publishing House, Moscow, p. 248 (Russian) (2008).
45. Ermilova, I. A., Alekseeva, L. N., Shamolina, I. I., and Khokhlova, V. A. Effect of microorganisms on the structure of synthetic fibers. *Tekstil'naya Promyshlennost' (Textile Industry) journal*, 9, 55–57(Russian) (2010).
46. Watanabe, T. and Miyazaki, K. *Morphological deterioration of acetate, acrylic, poliamide and polyester textiles by micro-organisms* (Aspergillus spp., Penicillium spp.). Sen.-1 Gakkaishi, 36, pp. 409–415 (2009).
47. Perepelkin, K. E. *Structure and properties of fibers.*, Khimiya (Chemistry) Publishing House, Moscow, p. 208 (Russian) (2009).
48. Mankriff, R. U. *Chemical fibers.* By A. B. Pakshver (Ed.), Legkayz Industriya (Light Industry) Publishing House, Moscow, p. 606 (Russian) (2011).
49. Kudryavtsev, G. I., Nosov, M. P., and Volokhina, A. V. *Polyimide fibers.* Khimiya (Chemistry) Publishing House, Moscow, p. 264 (Russian) (2010).
49. Tager, A. A. *Physico-chemistry of polymers.* Khimiya (Chemistry) Publishing House, Moscow, p. 544 (Russian) (2012).
50. *Fibers with special properties.* By L. A. Wolf (Ed.), Khimiya (Chemistry) Publishing House, Moscow, p. 240 (Russian) (2012).
51. Hamlyn, P. F. Microbiological deterioration of textiles. *Textiles*, 12(3) 73–76 (2011).
52. Hofman, H. P. Die antimikrobielle Ausrustung der Kleidung. *Textiltechnik,.* 36(1) S30–32 (2010).
53. Vigo, T. L. and Benjaminson, M. A., *Textile Research Journal* 51(7), 454–465 (2010).
54. Kozinda, Z. Yu., Gorbacheva, I. N., Suvorova, E. G., and Sukhova, L. M. *Obtaining methods of textile materials with specific (antimicrobes and flame retardants) properties.* Legprombytizdat (Light Industry) Publishing House, Moscow, p. 112 (Russian) (1988).
55. McCarthy, B. J. Rapid methods for the detecbion of biodeterioration in textiles. *International Biodeterioration*, 23, 357–364 (2009).
56. Mc Carthy, B. J. Preservatives for use in the Wool Textile Industry. *Preservatives in the Food, Pharmaceutical and Environmental Industries.* R. G. Board, M. C. Allwood, and J. G. Bauks (Eds.), Blackwell Scientific Publications, London, pp. 75–98 (1987).
57. *Anon. Preservative treatments for textiles. Part I. Specification for treatments.* British Standard 2087. British Standards Institution, London (1981).
58. Kalontarov, I. Ya. *Properties and methods of application of active pigments.* Donish Publishing House, Dushanbe, p. 126 (Russian) (1970).

59. Emanuel, N. M., Zaikov, G. E., and Maizus, Z. K. *Oxidation of organic compounds*. Medium effects in radical reactions. Pergamon Press, Oxford, p. 628 (1984).

11 New Application for Reuse of Thermoset Plastics

A. K. Haghi and G. E. Zaikov

CONTENTS

11.1 INTRODUCTION

The plastic materials production has reached global maximum capacities leveling at 260 million tones in 2007, where in 1990 the global production capacity was estimated at 80 million tones. It is estimated that production of plastics worldwide is growing at a rate of about 5% per year [1]. It is low density, strength, user-friendly designs, fabrication capabilities, long life, light weight, and low cost are the factors behind such phenomenal growth. The plastics also contribute to our daily life functions in many aspects. With such large and varying applications, plastics contribute to an ever increasing volume in the solid waste stream [2].

The major problems that this level of waste production generates initially entail storage and elimination. Storage implies the availability of large surface areas and their management, which in turn requires the installation of specific containers according to the nature of the products that are deposited there. Elimination is most typically carried out by either land filing or incineration (regardless of the type of plastic waste) and by regeneration (only in the case of thermoplastic products) [3]. The plastics can be separated into two types. The first type is thermoplastic, which can be melted for recycling in the plastic industry, such as polyethylene, polyamide and polyethylene terephthalate. The second type is thermosetting plastic. This plastic cannot be melted by heating because the molecular chains are bonded firmly with meshed cross-links such as unsaturated polyester, polyurethane, and melamine. At present, these plastic wastes are disposed by either burning or burying [4]. The burning causes to emissions of carbon-dioxide (CO_2), nitrogen oxide (NO), and sulfur-dioxide (SO_2). On the other hand, plastic materials remain in the environment for 100, perhaps 1,000 of years. Therefore, both the ways contributing to the environmental problems. Within this framework, the objective of this chapter is to find alternatives for managing the plastic wastes while protecting the environment. To achieve this purpose, a study of these thermosetting plastics has been recycled for use into construction materials.

However, a lightweight structure is also desirable in earthquake prone areas [5]. It is convenient to classify the various types of lightweight concrete by their method of production. These are [6]:

(a) By using porous lightweight aggregate of low apparent specific gravity, that is lower than 2.6, for example, pumice material. This type of concrete is known as lightweight aggregate concrete.

(b) By introducing large voids within the concrete or mortar mass; these voids should be clearly distinguished from the extremely fine voids produced by air entrainment such as aluminum powder. This type of concrete is variously known as aerated, cellular, and foamed or gas concrete.

(c) By omitting the fine aggregate from the mix so that a large number of interstitial voids are present; normal weight coarse aggregate is generally used. This concrete is known as no fines concrete.

11.2 BACKGROUND

The possible use of recycled plastic waste in concrete and other construction materials has been studied by a number of researchers. It was reported by Naik et al. [7] that compressive strength decreased with an increase in the amount of the plastic in concrete, particularly above 0.5% plastic addition to total weight of the mixture. Choi et al. [8] investigated the quality of lightweight aggregates, conducting tests on the workability and the strength properties of concrete, analyzing the relationship between the quality of aggregates and the properties of concrete. Lightweight aggregates were made from waste PET bottles, and granulated blast-furnace slag (GBFS) was used to examine whether it is possible to improve the quality of lightweight aggregate. The 28 day compressive strength of waste PET bottles lightweight aggregate concrete (WPLAC) with the replacement ratio of 75% reduces about 33% compared to the control concrete in the water–cement ratio of 45%. The density of WPLAC varies from 1,940 to 2,260 kg/m³ by the influence of waste PET bottles lightweight aggregate (WPLA). The structural efficiency of WPLAC decreases as the replacement ratio increases. The workability of concrete with 75% WPLA improves about 123% compared to that of the normal concrete in the water–cement ratio of 53%. The adhered GBFS is able to strengthen the surface of WPLA and to narrow the transition zone owing to the reaction with calcium hydroxide. Marzouk et al. [3] studied the effects of PET waste on the density and compressive strength of concrete. It was found that the density and compressive strength decreased when the PET aggregates exceeded 50% by volume of sand. The density and compressive strength of concrete were between 1,000 and 2,000 kg/m³ and 5-60 MPa, respectively. Batayneh et al. [9] investigated the performance of the ordinary portland cement (OPC) concrete mix under the effect of using recycled waste materials, namely glass, plastics, and crushed concrete as a fraction of the aggregates used in the mix. The main findings of this investigation revealed that the three types of waste materials could be reused successfully as partial substitutes for sand or coarse aggregates in concrete mixtures. Similarly, the results of Ismail and AL-Hashmi [10] showed that reusing waste plastic as a sand-substitution aggregate in concrete gives a good approach to reduce the cost of materials and solve some of the solid waste problems posed by plastics. Recently, Albano et al. [11] and Choi et al. [5] also studied the mechanical behavior of concrete with recycled PET, varying the PET content. Results indicate that when volume proportion of PET increased in concrete, showed a decrease in compressive strength, however, the water absorption increased.

The main objective of this study is to investigate the possibility of improvement results of the studied by Panyakapo, P. and Panyakapo, M. [4] that used of melamine as thermosetting plastics for application into construction materials has been conducted, particularly for the concrete wall in buildings.

11.3 MATERIALS

The materials used in present study are as follows:

11.3.1 Ordinary Portland Cement
Type I Portland cement conforming to ASTM C150 [12].

11.3.2 Sand

Fine aggregate is taken from natural sand. Table 1 represents the properties of the sand and its gradation is shown in Figure 1.

TABLE 1   Properties of sand and melamine aggregates.

Properties	Sand	Melamine
density (g/cm3) [4]	2.60	1.574
bulk density (g/cm³) [13]	–	0.3-0.6
Water absorption (%)	1.64	7.2
Max size (mm)	4.75	1.77
Min size (mm)	–	0.45
Sieve 200 (%)	0.24	–
Flammability [13]	non flammable	non flammable
Decomposition [13]	–	at > 280°C formation of NH_3

11.3.3   Thermosetting Plastic

Melamine is a widely used type of thermosetting plastic. Therefore, in the present work has been selected for application in the mixed design of composite. The mechanical and physical properties of melamine are shown in Table 1. The melamine waste was ground with a grinding machine. The ground melamine waste was separated under sieve analysis. The results of cumulative percentage passing from a set of sieves were compared with the grading requirements for fine aggregate according to ASTM C33 [14]. Scanning electron micrographs (SEM) of melamine aggregates is shown in Figure 2. Panyakapo, P. and Panyakapo, M. [4] reported that in the process of producing lightweight concrete, aerated concrete was produced by the formation of gas, which rises to the surface of concrete. Therefore, the very small plastic particles tended to rise to the concrete surface along with the rising gas and the relatively large plastic particles segregated to the bottom part of concrete. To avoid the segregation of fine and coarse particles and protect the gradation curve of the combination of sand and plastic aggregates in limited of the requirements of ASTM C33 [14], the appropriate particle sizes were chosen from those retained on sieve numbers 10-40.

The melamine aggregates were been saturated surface dry. Therefore, melamine aggregates immerse in water at approximately 21°C for 24 hr and removing surface moisture by warm air bopping.

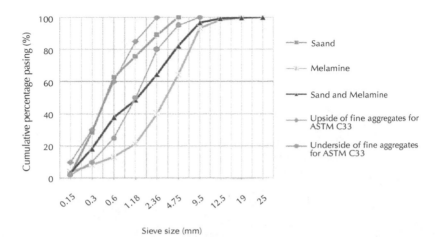

FIGURE 1 Gap grading analysis of sand and melamine aggregates according to ASTM C33 [14].

(a)

FIGURE 2 *(Continued)*

(b)

FIGURE 2   The SEM photographs of materials: (a) melamine aggregates and (b) silica fume.

11.3.4 Aluminum Powder

In the present study, aluminum powder was selected as an agent to produce hydrogen gas (air entrainment) in the cement. This type of lightweight concrete is then called aerated concrete. The following are possible chemical reactions of aluminum with water:

$$2Al + 6H_2O \rightarrow 2Al(OH)_3 + 3H_2 \tag{1}$$

$$2Al + 4H_2O \rightarrow 2AlO(OH) + 3H_2 \tag{2}$$

$$2Al + 3H_2O \rightarrow Al_2O_3 + 3H_2 \tag{3}$$

The first reaction forms the aluminum hydroxide bayerite ($Al(OH)_3$) and hydrogen, the second reaction forms the aluminum hydroxide boehmite ($AlO(OH)$) and hydrogen, and the third reaction forms aluminum oxide and hydrogen. All these reactions are thermodynamically favorable from room temperature past the melting point of aluminum (660°C). All are also highly exothermic. From room temperature to 280°C, $Al(OH)_3$ is the most stable product, while from 280-480°C, $AlO(OH)$ is most stable. Above 480°C, Al_2O_3 is the most stable product (3) ([15, 16]). The following equation illustrates the combined effect of hydrolysis and hydration on tricalcium silicat.

$$3CaO.SiO_2 + water \rightarrow xCaO.ySiO_2(aq.) + Ca(OH) \qquad (4)$$

In considering the hydration of Portland cement it is demonstrate that the more basic calcium silicates are hydrolysis to less basic silicates with the formation of calcium hydroxide or 'slaked lime' as a by-product. It is this lime which reacts with the aluminum powder to form hydrogen in the making of aerated concrete from Portland cement [17]:

$$2Al + 3Ca(OH)_2 + 6H_2O \rightarrow 3CaO.Al_2O_3.6H_2O + 3H_2 \qquad (5)$$

Hydrogen gas creates many small air (hydrogen gas) bubbles in the cement paste. The density of concrete becomes lower than the normal weight concrete due to this air entrainment.

11.3.5 Silica Fume
In the present work, Silica fume has been used. It is chemical compositions and physical properties are being given in Table 2 and Table 3, respectively. The SEM of silica fume is shown in Figure 2.

11.3.6 Superplasticize
$r:$ Premia 196 with a density of 1.055 ± 0.010 kg/m^3 was used. It was based on modified polycarboxylate.

TABLE 2 Chemical composition of silica fume.

Chemical composition	Silica fume
SiO_2 (%)	86–94
Al_2O_3 (%)	0.2–2
Fe_2O_3 (%)	0.2–2.5
C (%)	0.4–1.3
Na_2O (%)	0.2–1.5
K_2O (%)	0.5–3
MgO (%)	0.3–3.5
S (%)	0.1–0.3
CaO (%)	0.1–0.7
Mn (%)	0.1–0.2
SiC (%)	0.1–0.8

TABLE 3　Physical properties of silica fume.

Items	Silica fume
	2.2–2.3
particle size (μm)	< 1
Specific surface area (m²/gr)	15–30
Melting point (°C)	1230
Structure	amorphous

11.4　MIX DESIGN

To determine the suitable composition of each material, the mixing proportions were tested in the laboratory, as shown in Table 4. In this study, the mix proportions were separated for five experimental sets. For each set, the cement and Aluminum powder contents was specified as a constant proportion. The proportion of each of the remaining materials, that is sand, water, silica fume, aluminum powder, and melamine, was varied for each mix design.

For mix number 1, to determination the primal proportion of melamine plastic content in the composition. The concrete was composed of cement, sand, melamine, and plastic, adjusted to 1.0:1.0:1.0-3.0. The plastic proportion was increased in increments of 0.5. The water and aluminum powder were taken as 0.35 and 0.004 by weight of cement, respectively.

For mix number 2, to determine the optimum proportion of sand, the quantities of sand by weight of cement were varied from 1.0 to 1.8 with an increment of 0.2 for each step. The portions of other materials were kept constant. That is the proportion of cement, aluminum powder, water, and melamine plastic was adjusted to 1.0:0.004:0.35:1.0.

For mix number 3, to determine the optimum water content in the composition, the quantities of water were varied from 0.3 to 0.55 by an increment of 0.05 for each step. The proportions of other materials were kept constant. That is the composition of cement, aluminum powder, sand, and melamine plastic was adjusted to 1.0:0.004:1.4:1.0.

For mix number 4, to determine the optimum proportion of silica fume content in the composition, the quantities of silica fume were varied from 0.1 to 0.35 by an increment of 0.05 for each step. Due to the reduced workability of the concrete containing silica fume, superplasticizer should be used. It was used to control the slump. Other materials contents were kept the same as the previous tests. That is, the composition of cement, aluminum powder, sand, water, and melamine plastic was adjusted to 1.0: 0.004:1.4:0.35:1.0.

For the last mix number, to determine the final proportion of melamine plastic content in the composition, the quantities of melamine plastic were varied from 1.0 to 2.2 by an increment of 0.2 for each step. Superplasticizer was used to control the slump. The proportions of other materials were kept constant. That is the composition

TABLE 4 Mix proportions of melamine lightweight composites (by weight).

Mix no.	Cement	Aluminum powder	Sand	Water	Silica fume	Melamine	Superplasticizer
1. Determination of melamine content (1st trial mix design)	1.0	0.004	1.0	0.35	—	1.0	—
						1.5	
						2.0	
						2.5	
						3.0	
2. Determination of sand content	1.0	0.004	1.0	0.35	—	1.0	—
			1.2				
			1.4				
			1.6				
			1.8				
3. Determination of water content or water–cement ratio (w/c)	1.0	0.004	1.4	0.30	—	1.0	—
				0.35			
				0.40			
				0.45			
				0.50			
				0.55			

TABLE 4 *(Continued)*

Mix no.	Cement	Aluminum powder	Sand	Water	Silica fume	Melamine	Superplasticizer
4. Determination of silica fume content	1.0	0.004	1.4	0.35	0.10	1.0	0.005
					0.15		0.007
					0.20		0.009
					0.25		0.012
					0.30		0.015
					0.35		0.020
5. Determination of melamine content (final mix design)	1.0	0.004	1.4	0.35	0.25	1.0	0.012
						1.2	0.011
						1.4	0.010
						1.6	0.009
						1.8	0.008
						2.0	0.007
						2.2	0.006

of cement, aluminum powder, sand, water, silica fume, and melamine was adjusted to 1.0:1.4:0.004:0.35:0.25:1.0-2.2.

11.5   EXPERIMENTAL TECHNIQUES

Mortar was mixed in a standard mixer and placed in the standard mold of $50 \times 50 \times 50$ mm according to ASTM C109 [18]. In the pouring process of mortar, an expansion of volume due to the aluminum powder reaction had to be considered. The expanded portion of mortar was removed until finishing. The fresh mortar was tested for slump according to ASTM C143 [19]. The specimens were cured by wet curing at normal room temperature. The hardened mortar was tested for dry density, compressive strength, water absorption, and voids for the curing age of 7 and 28 days. The test results for melamine, sand and water contents were reported for 7 days curing age for mix nos. 1-3, because these were very close to the results of 28 days. When silica fume was added in the latter mix nos. 4 and 5, the test results were presented for 28 days. This is because the presence of silica fume increases the duration for completion of the chemical reaction. The testing procedures of dry density, water absorption, and voids were performed according to ASTM C642 [20] and compressive strength was performed according to ASTM C109 [18].

11.6   DISCUSSION AND RESULTS

The Figure 3 present the variations in the compressive strength and dry density for 7 days age of mortars as a function of the value of melamine substitutes used. It can initially be seen, to increased melamine, the compressive strength and dry density of composites decreased. It was found that the highest compressive strength (3.23 MPa) was obtained for the ratio of cement, aluminum powder, water, sand, and melamine of 1.0:0.004:0.35:1.0:1.0. However, the compressive strength for this mix proportion does not exactly satisfy the standard value. Table 5 shows that the specification of non-load-bearing lightweight concrete according to ASTM C129 [21] Type II.

The reduction in the compressive strength due to the addition of melamine aggregates might be due to either a poor bond between the cement paste and the melamine aggregates or to the low strength that is characteristic of plastic aggregates. Due to this reason, the proportion of melamine plastic equal to 1.0 was selected for the next mix design. After the rest of the composition was determined, the melamine plastic was tested again to find the suitable proportion.

TABLE 5   Specification of non-load-bearing lightweight concrete [21].

Type	Compressive strength (MPa)		Density (kg/m³)
	Average of three unit	Individual unit	
II	4.1	3.5	< 1680

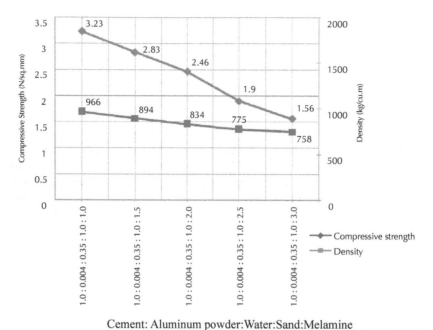

Cement: Aluminum powder:Water:Sand:Melamine

FIGURE 3 Compressive strength and density for varying melamine content (curing for 7 days).

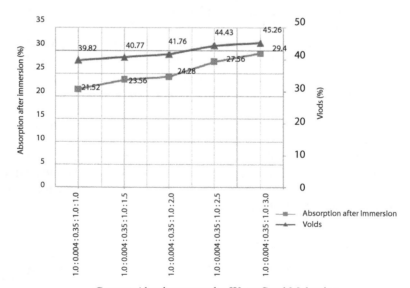

Cement:Aluminum powder:Water:Sand:Melamine

FIGURE 4 Absorption after immersion for varying melamine content (curing for 7 days).

The absorption is an indirect parameter to examine the inside porosity of mortar. The results showed that the absorption after immersion and voids of mortar increased as the melamine content increased (see Figure 4 and Figure 5). Therefore, to increased melamine plastic, the inside porosity of mortar increased. This might be other reason for the reduction in the compressive strength and density.

FIGURE 5 Optical photographs of samples containing varying melamine, right to left containing 1.0, 1.5, 2.0, 2.5, and 3.0 the weight percentage of melamine.

11.6.1 Mix Number 2 (Determination of Sand Content)

The results of compressive strength and dry density for 7 days age are shown in Figure 6. It can be seen that a reduction of sand leads to a reduction in the strength and dry density. The compressive strength and dry density for sand content equal to or greater than 1.4 exactly satisfy the standard value. The proportion of sand equal to 1.4 was selected for the next mix design and the resumed optimize of material. Because the ultimate aim of this paper was to determine the suitable proportion to achieve the lowest dry density and acceptable compressive strength for non-load-bearing lightweight concrete according to ASTM C129 Type II standard.

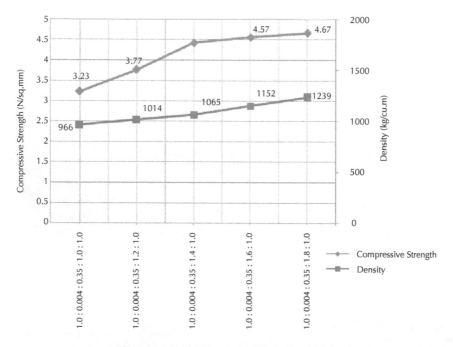

Cement:Aluminum powder:Water:Sand:Melamine

FIGURE 6 Compressive strength and density for the determination of the optimum sand content (curing for 7 days).

The Figure 7 presents the variations in the absorption after immersion and voids as a function of the value of sand substitutes used. The results showed that the absorption after immersion and voids of mortar decreased as the sand content increased. It can be concluded that, the inside porosity of mortar was decreased, when the sand content increased in the mix. The reduction in the absorption after immersion and voids due to the addition of sand might be due to either a good bond between the cement paste and the sand aggregates or to the sand aggregates fine than the melamine aggregates.

11.6.2 Mix Number 3 (Determination of Water Content)

The results of compressive strength and dry density for 7 days age are shown in Figure 8. The results showed that the compressive strength and dry density of mortar decreased as the water content increased.

The results of the absorption after immersion and voids tests for varying water content are illustrated in Figure 9. The results showed that the absorption after immersion and voids of mortar increased as the water content increased. It was reported by Choi et al [8] and Albano et al. [11] that, when the w/c ratios are higher, there is an excess of water that does not participate in the water-cement reaction, so channels with very small diameters like capillaries are produced and when the water evaporates, those empty spaces rest resistance to the concrete. It can be the important reason for decrease of compressive strength and density and increase of absorption after immersion and voids of mortar. It was found that the optimum water content, which leads to the maximum strength, is equal to 0.30. But to achieve the lowest dry density and acceptable compressive strength, the proportion of water equal to 0.35 was selected for the next mix design.

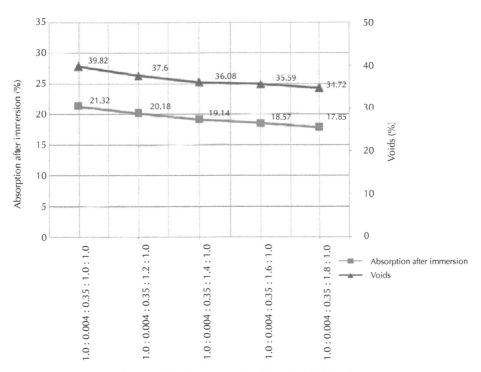

Cement: Aluminum powder:Water:Sand:Melamine

FIGURE 7   Absorption after immersion and voids for varying sand content (curing for 7 days).

11.6.3   Mix Number 4 (Determination of Silica Fume Content)

The Figure 10 present the variations in the compressive strength and dry density for 28 days age of mortars as a function of the value of silica fume substitutes used. It was found that the results of compressive strength for 7 days age do not increase when compared with those without silica fume. This may be due to the incomplete chemical. The curing age extended to 28 days for the complete chemical reaction between silica fume and water. The figure shows that the compressive strength and dry density of mortar increased as the silica fume content increased to 0.25. But for high silica fume content greater than 0.25, the values of compressive strength decreased. When considering the use of SF as an addition, the micro filling effect and pozzolanic reaction of SF contributed to a denser microstructure thus resulting in an increase in the compressive strength. A similar result was reported by Rao G.A. [22] for the compressive strength of silica fume concrete. An optimum silica fume content of 0.25 was selected as the suitable proportion. The value of compressive strength was equal to 12.06 MPa for silica fume content of 0.25. It was very more than the value of the average of three units (4.1 MPa) as specified for non-load-bearing lightweight concrete according to ASTM C129 Type II standard. Therefore, to achieve the lowest dry density and acceptable compressive strength the used more melamine plastic in concrete for the next mix designs.

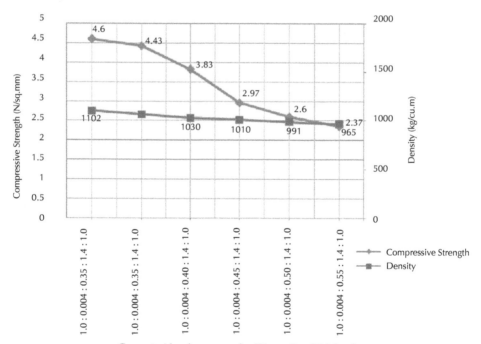

FIGURE 8 Compressive strength and density for the determination of the optimum water content (curing for 7 days).

The results of the absorption after immersion and voids tests are shown in Figure 11. It can be seen that the replacement of weight material by silica fume can effectively reduce the absorption and voids. For silica fume concrete, the higher the replacement level, the more the reducing effect on absorption and voids. This reduction is due to the filling effect of SF contributed to a denser microstructure. A similar result was reported by Chan and Ji [23] for the silica fume concrete.

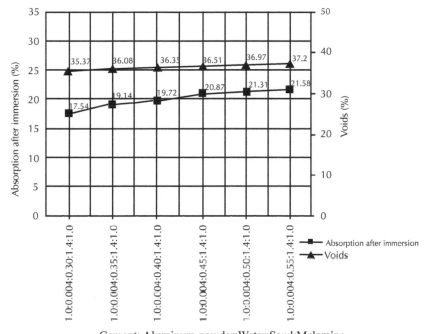

Cement: Aluminum powder:Water:Sand:Melamine

FIGURE 9 Absorption after immersion and voids for varying water content (curing for 7 days).

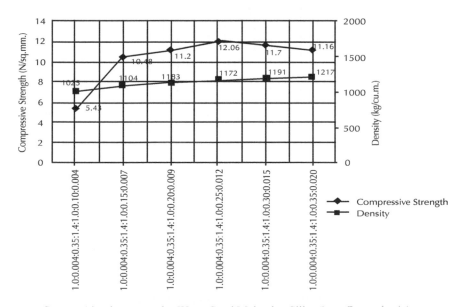

Cement:Aluminum powder:Water:Sand:Melamine:Silica fume:Superplasticizer

FIGURE 10 Compressive strength and density for the determination of the optimum silica fume content (curing for 28 days).

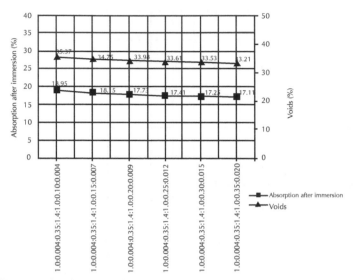

Cement: Aluminum powder:Water:Sand:Melamine:Silica fume:Superplasticizer

FIGURE 11   Absorption after immersion and voids for varying silica fume content (curing for 28 days).

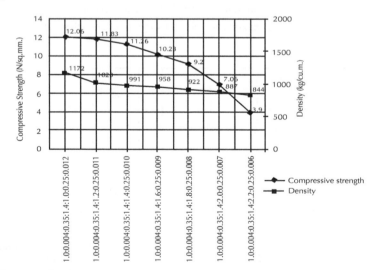

Cement: Aluminum powder:Water:Sand:Melamine:Silica fume:Superplasticizer

FIGURE 12   Compressive strength and density for the determination of the optimum melamine plastic content (curing for 28 days).

11.6.4   Mix Number 5 (Determination of the Final Melamine Plastic Content)

The Figure 12 present the variations in the compressive strength and dry density for 28 days age of mortars as a function of the value of melamine substitutes used. It was found that the presence of melamine caused a reduction in the dry density and compressive strength of concretes. Figure 13 show that the scanning electron microscopy analysis of composites reveals that cement paste melamine aggregates adhesion is imperfect and weak. Therefore, the problem of bonding between plastic particles and cement paste is main reason to decrease of compressive strength. An optimum melamine content of 2.0 was selected. The results of compressive strength and dry density, which are 7.06 MPa and 887 kg/m³, are according to ASTM C129 Type II standard.

The results showed that the absorption after immersion and voids of mortar increased as the melamine content increased (see Figure 14). Also, the structure analysis of mortars by scanning electron microscopy has revealed a low level of compactness in mortars when the value of melamine plastic increased (see Figure 15). It was confirmed that to increased melamine plastic, the inside porosity of mortar increased.

11.6.5 Comparison

Based on the results, the optimum proportions of materials are cement:aluminum powder:water:sand:melamine:silica fume:superplasticizer equal to 1.0:0.004:0.35: 1.4:2.0:0.25:0.007. However, the optimum proportions of materials for the studied by [4] were cement, sand, water, fly ash, aluminum powder, and melamine equal to 1.0:0.8:0.75:0.3:0.0035:0.9. Therefore, the used of melamine plastic more than previous study (about 122%) for non-load-bearing lightweight concrete according to ASTM C129 Type II standard. Also, the dry density of non-load-bearing lightweight concrete was reduction by 36.4% in comparison with the previous study.

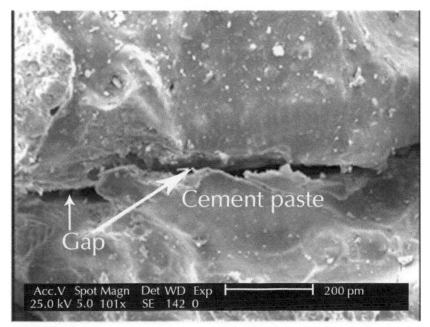

FIGURE 13 Microstructure of concrete containing 1.6 melamine by weight of cement, as obtained using SEM (enlargement: 101×).

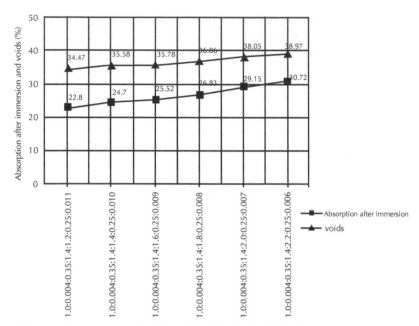

Cement:Aluminum powder:Water:Sand:Melamine:Silica fume:Superplasticizer

FIGURE 14   Absorption after immersion and voids for varying melamine plastic content (curing for 28 days).

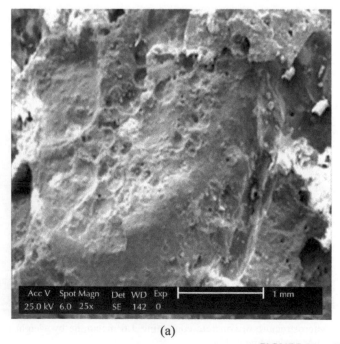

(a)

FIGURE 15   *(Continued)*

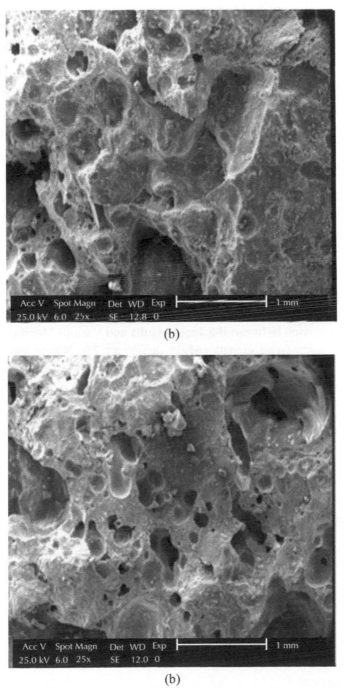

(b)

(b)

FIGURE 15 SEM of various mortars containing melamine plastic aggregates (a) 1.2 by weight of cement (enlargement: 25×), (b) 1.6 by weight of cement (enlargement: 25×), and (c) 2.0 by weight of cement (enlargement: 25×).

11.6.6 Comparison Research Findings on the Use of Waste Plastic in Concrete

The results of this study are in a perfectly agreement with the other research findings on the use of waste plastic in concrete (see Table 6).

TABLE 6 Comparsion between results of this study and other research findings on the use of waste plastic in concrete (by increase plastic).

Description	Type of waste plastic	Compressive strength	Density	Absorption	void
This study	Melamine	Decrease	Decrease	Increase	Increase
[11]	PET	Decrease	–	Increase	Increase
	PVC	Decrease	Decrease	–	–
[8]	PET	Decrease	Decrease	Increase	Increase
[4]	Melamine	Decrease	Decrease	–	–
[10]	80% polyethylene and 20% polystyrene	Decrease	Decrease	–	–
[3]	PET	Decrease	Decrease	–	–

11.6.7 Comparison between the Tests Results and Various Standards

The results of compressive strength and dry density are 7.06 N/mm² and 887 kg/m³, respectively. It was found that the compressive strength of plastic lightweight concrete exactly satisfy the class 4 of aerated lightweight concrete according to Institute of Standards and Industrial Research of Iran (ISIRI 8593-1st.Edition). However, the dry density is some greater than the ISIRI standard. In addition, the results of this study exactly satisfy the specifications of rendering or plastering mortar for utilization of public and rendering or plastering lightweight mortar according to ISIRI 706-1 standard (1st. Revision). Rendering and plastering mortar are widely used in Iran. These comparisons are summarized in Table 7.

TABLE 7 Comparison between the test results of this study and various standards.

Description	Compressive strength (N/mm²) (Average of three units)	Dry density (kg/cm³)
Plastic lightweight concrete (this study), cement: aluminum powder : water : sand : melamine: silica fume : superplasticizer = 1.0:0.004:0.35:1.4:2.0:0.25:0.007	7.06	887
Aerated concrete masonry units, Class 4 (ISIRI 8593-1st.Edition)	5.0	450 - 860
Rendering or plastering mortar for utilization of public (ISIRI 706-1,1st.Revision)	0.4 <	–
Rendering or plastering lightweight mortar (ISIRI 706-1,1st.Revision)	0.4 - 7.5	≤ 1300

11.7 CONCLUSION

Melamine plastic aggregates can be successfully and effectively utilized for non-load-bearing lightweight concrete according to ASTM C129 Type II standard. The following conclusions were drawn from the investigation:

1. With an increase of replacement ratio of materials of lightweight concrete by melamine plastic aggregates:

    (a) The compressive strength and densities of the lightweight concrete were reduced. The reduction in the compressive strength due to the addition of melamine aggregates might be due to either a poor bond between the cement paste and the melamine aggregates or to the low strength that is characteristic of plastic aggregates. Also, to increased melamine plastic, the inside porosity of mortar increased. This might be other reason for the reduction in the compressive strength and density. The other reason for the reduction in the density is due to low specific gravity of melamine plastic than sand.

    (b) The absorption and volume of permeable voids of the lightweight concrete were increased.

2. The existed straight relationship between the compressive strength and density. Also, the existed inverse relationship between the compressive strength, density and absorption, volume of permeable voids.

3. Density and compressive strength of mortar decreased as the water content increased. Because, when the w/c ratios are higher, there is an excess of water that does not participate in the water-cement reaction, so channels with very small diameters like capillaries are produced and when the water evaporates, those empty spaces rest resistance to the concrete.

4. The compressive strength increase as the addition of SF increases. However, the dry density tends to also increase due to the relatively high specific gravity. Also, absorption and volume of permeable voids decrease as the addition of SF increases. These happen due to the micro filling effect and pozzolanic reaction of SF contributed to a denser microstructure.

5. The scanning electron microscopy analysis of composites reveals that cement paste melamine aggregates adhesion is imperfect and weak. Also, The scanning electron microscopy shows that with an increase of replacement ratio of materials of lightweight concrete by melamine plastic aggregates, the pores and cavernous in the structure of the lightweight concrete were increased.

Utilization of other plastics in the mix proportion for non-load-bearing lightweight concrete according to ASTM C129 Type II standard and other standards are suggest for further studies.

KEYWORDS

- Aluminum powder
- Granulated blast-furnace slag
- Ordinary portland cement
- Plastic materials
- Scanning electron micrographs

REFERENCES

1. Al-Salem, S. M., Lettieri, P., and Baeyens, J. The valorization of plastic solid waste (PSW) by primary to quaternary routes: From re-use to energy and chemicals. *Progress in Energy and Combustion Science*, **36**, 103-129 (2010).
2. Siddique, R., Khatib, J., and Kaur, I. Use of recycled plastic in concrete: A review. *Waste Management*, **28**, 1835–1852 (2008).
3. Marzouk, O. Y., Dheilly, R. M., and Queneudec, M. Valorization of post-consumer waste plastic in cementitious concrete composites. *Waste Management*, **27**, 310–318 (2007).
4. Panyakapo, P. and Panyakapo, M. Reuse of thermosetting plastic waste for lightweight concrete. *Waste Management*, **28**, 1581–1588 (2008).
5. Choi, Y. W., Moon, D. J., Kim, Y. J., and Lachemi, M. Characteristics of mortar and concrete containing fine aggregate manufactured from recycled waste polyethylene terephthalate bottles. *Construction and Building Materials*, **23**, 2829–2835 (2009).
6. Neville, A. M. and Brooks, J. J. *Concrete Technology*. Longman Scientific and Technology co published in the US with John Wiley and Sons, New York, p. 345 (1991).
7. Naik, T. R., Singh, S. S., Huber, C. O., and Brodersen, B. S. Use of postconsumer waste plastics in cement-based composites. *Cement and Concrete Research*, **26**(10), 1489–1492 (1996).
8. Choi, Y. W., Moon, D. J., Chung, J. S., and Cho, S. K. Effects of waste PET bottles aggregate on the properties of concrete. *Cement and Concrete Research*, **35**, 776–781 (2005).
9. Batayneh, M., Marie, I., and Asi, I. Use of selected waste materials in concrete mixes. *Waste Management*, **27**, 1870–1876 (2007).
10. Ismail, Z. Z. and AL-Hashmi, E. A. Use of waste plastic in concrete mixture as aggregate replacement. *Waste Management*, **28**, 2041–2047 (2008).
11. Albano, C., Camacho, N., Hernandez, M., Matheus, A., and Gutierrez, A. Influence of content and particle size of waste pet bottles on concrete behavior at different w/c ratios. *Waste Management*, **29**, 2707–2716 (2009).
12. *Annual Book of ASTM Standard*. Specification for Portland cement, ASTM C150 (1994).
13. Melamine, CAS N:108-78-1. Unep Publications.
14. *Annual Book of ASTM Standard*. Specification for Fine and Coarse Aggregates, ASTM C33 (1992).
15. U. S. Department of Energy. Reaction of Aluminum with Water to Produce Hydrogen. Version 1.0. pp. 1–26 (2008).
16. Studart, A. R., Innocentini, M. D. M., Oliveira, I. R., and Pandolfelli, V. C. Reaction of aluminum powder with water in cement-containing refractory castables. *Journal of the European Ceramic Society*, **25**, 3135–3143 (2005).
17. Short, A. and Kinniburgh, W. *Lightweight Concrete*. Applied Science publishers LTD London, Third Edition, p. 18 (1978).
18. *Annual Book of ASTM Standard*. Standard Test Method for Compressive Strength of Hydraulic Cement Mortars, ASTM C109 (2002).
19. *Annual Book of ASTM Standard*. Test Method for Slump of Hydraulic—Cement Concrete, ASTM C143 (2003)

20. *Annual Book of ASTM Standard*. Test Method for Density, Absorption, and Voids in Hardened Concrete, ASTM C642 (1997)
21. *Annual Book of ASTM Standard*. Standard Specification for Nonloadbearing Concrete Masonry Units, ASTM C129 (2005).
22. Rao, G. A. Investigations on the performance of silica fume incorporated cement pastes and mortars. *Cement and Concrete Research*, **33** 1765–1770 (2003).
23. Chan, Y. N. S. and Ji, X. Comparative study of the initial surface absorption and chloride diffusion of high performance zeolite, silica fume and PFA concretes. *Cement and Concrete Composites*, **21** 293–300 (1999).

Index